Plumbing

Plumbing

Alan Wakeford

Consultant Editor Richard Wiles

OCTOPUS BOOKS

This edition published in 1989
by Octopus Books Limited,
a division of the Octopus Publishing Group,
Michelin House,
81 Fulham Road,
London SW3 6RB

Copyright © Hennerwood Publications Limited, 1984

ISBN 0 7064 2958 3

Produced by Mandarin Offset
Printed and bound in Hong Kong

Contents

6 DOMESTIC PLUMBING EXPLAINED

How often do you think about the plumbing system in your house? Every day you use water for drinking and cooking, bathing and showering, and to flush WC cisterns, but for the most part because the system is out of sight it's usually out of mind as well. Indeed, a properly installed system will operate trouble-free for years, so why shouldn't you take it for granted that when you turn on a tap, water will come out of it?

This is not to say that you can ignore the plumbing totally, however. To work well, it does require some maintenance; leaking taps occasionally have to be dealt with, for instance, as do blocked drains. Emergencies, too, can arise unexpectedly, and sooner or later you may well want to make some modifications to the system.

There's something of a mystique about a plumbing system, but although the labyrinth of pipes may seem complicated, in reality it isn't. If something needs to be done, your first reaction may be to call in the plumber; however, with a basic knowledge of what's going on and a few practised skills, you'll be able to deal with most problems as they occur and to carry out a

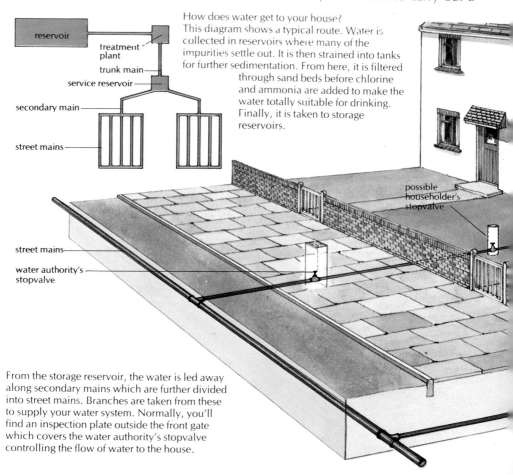

How does water get to your house? This diagram shows a typical route. Water is collected in reservoirs where many of the impurities settle out. It is then strained into tanks for further sedimentation. From here, it is filtered through sand beds before chlorine and ammonia are added to make the water totally suitable for drinking. Finally, it is taken to storage reservoirs.

reservoir

treatment plant

trunk main

service reservoir

secondary main

street mains

possible householder's stopvalve

street mains

water authority's stopvalve

From the storage reservoir, the water is led away along secondary mains which are further divided into street mains. Branches are taken from these to supply your water system. Normally, you'll find an inspection plate outside the front gate which covers the water authority's stopvalve controlling the flow of water to the house.

DOMESTIC PLUMBING EXPLAINED

range of plumbing projects. You might eventually want to plumb in a new washing machine or bathroom suite, or replace a cold water cistern — each is well within your scope and nowadays made easier because modern fittings and copper or plastic pipe are light and easy to use. If you're prepared to take the time with plumbing projects, you can save a considerable amount of money in labour costs.

Normally, if you're replacing appliances and parts of the plumbing but you're not changing their positions, you won't need to obtain the relevant Building Regulations approval. But if you want to install a completely new handbasin, for example, there are regulations that you should take into account. These are pointed out later, but it is still worth contacting your local building inspector and local water authority to seek their guidance regarding your particular system.

Usually, there's little problem in supplying water to an appliance; what you have to be careful of is how you get rid of the waste, and the regulations are there to make sure this is done hygienically and efficiently. Don't ignore them.

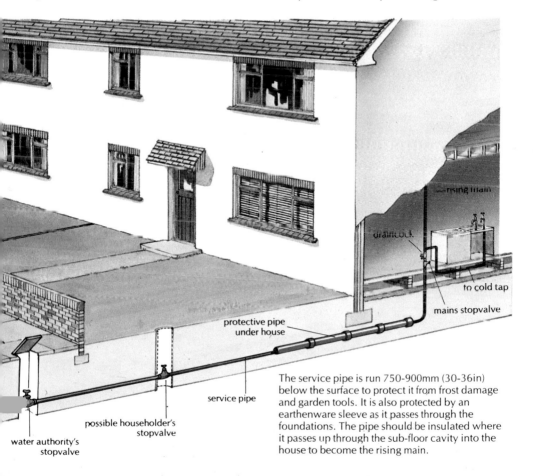

rising main

draincock

to cold tap

mains stopvalve

protective pipe
under house

service pipe

possible householder's
stopvalve

water authority's
stopvalve

The service pipe is run 750-900mm (30-36in) below the surface to protect it from frost damage and garden tools. It is also protected by an earthenware sleeve as it passes through the foundations. The pipe should be insulated where it passes up through the sub-floor cavity into the house to become the rising main.

8 HOW WATER REACHES YOUR HOUSE

The local water authority is responsible for supplying your house with water and for the disposal of the waste and sewage. What happens on your property is your responsibility, although your plumbing system must still conform to regulations and byelaws.

The local water supply is distributed in a series of trunk mains, mains and submains — pipes that run underneath the street — and branches are taken from these to supply individual houses. Naturally, the water authority has ultimate control over the flow to your system by means of an isolating stopvalve situated at the foot of a guard pipe just outside your house. You'll recognise it by its metal cover, usually set in the pavement. Only the water authority should operate the valve.

From this valve onwards, the plumbing becomes your responsibility. Water reaches the house along what is now called the service pipe. Near to where it enters the house, there should be another stopvalve. You may find it under the kitchen sink or even under the stairs. It is vital to know where this valve is situated because it will enable you to turn off the domestic supply in an emergency.

Such a stopvalve has a habit of jamming open because it is operated so infrequently; it is sensible, therefore, to turn it off and on a couple of times a year. If you turn it back half a turn from

fully open, it will be less prone to jamming, and won't affect the supply.

The stopvalve is where the plumbing really begins. Most houses have what is termed an 'indirect' cold supply, and at the hub of this system is a cold water storage cistern. Usually, it is sited in the loft and is supplied by the rising main, the pipe which continues from the service pipe. From here, water is fed under gravity pressure to the taps and WC cistern in the bathroom, and to the hot water cylinder. Water that has passed through the cistern is not considered to be suitable for drinking; the tap over the kitchen sink, then, is supplied direct from the rising main.

This type of system has a number of advantages, the most important being that there's a reservoir of water available — the cistern — for use if the mains supply is temporarily cut off. It also means that domestic water is supplied under an even but lower pressure than the mains, so reducing wear and tear on the system.

The alternative, largely outmoded set-up is known as a 'direct feed' system, whereby all the taps and WC cisterns are fed direct from the rising main, so they are all under mains pressure. However, this system still needs a cold water storage cistern to supply the hot water cylinder, and therefore the hot water pipes will be under gravity pressure.

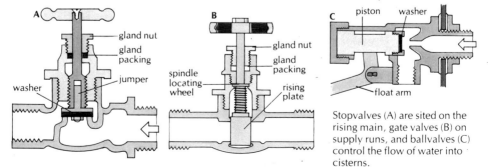

Stopvalves (A) are sited on the rising main, gate valves (B) on supply runs, and ballvalves (C) control the flow of water into cisterns.

HOW WATER REACHES YOUR HOUSE 9

The cold water system

Most of the cold taps and appliances in your home take their cold water from the main storage cistern. As the water flows under gravity pressure, the cistern must be higher than any of the outlets it supplies. Usually, it is situated in the loft, although it can be incorporated with a hot cylinder in what is termed a 'packaged' unit. The cistern itself is fed by the rising main which passes up through the house.

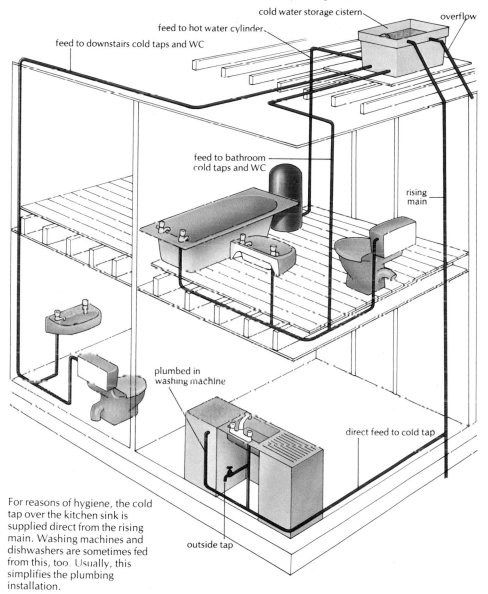

cold water storage cistern

overflow

feed to hot water cylinder

feed to downstairs cold taps and WC

feed to bathroom cold taps and WC

rising main

plumbed in washing machine

direct feed to cold tap

For reasons of hygiene, the cold tap over the kitchen sink is supplied direct from the rising main. Washing machines and dishwashers are sometimes fed from this, too. Usually, this simplifies the plumbing installation.

outside tap

10 HOT WATER SYSTEMS

Hot water at the turn of a tap whenever you want it is no longer considered a luxury — it's a necessity. There are now several ways in which this can be achieved, the simplest being an electric or gas instantaneous water heater (see page 67).

However, as with cold water systems, the commonest means of providing hot water is known as an indirect system. It revolves round a central heat source — a boiler or an immersion heater — and a storage cylinder which is fed from the cold water cistern. In the basic system, one where there's no combined central heating, a direct cylinder is used; water is drawn off from the bottom of the cylinder, fed to a boiler and then returned to the cylinder, where it's drawn off at the top

to supply the hot taps round the house. Or, there can be an immersion heater — rather like an electric kettle element — situated in the cylinder.

However, it is now more common to use an indirect cylinder, which is essential with a hot water central heating system. The cylinder incorporates a heat exchanger — a sort of 'radiator' made from a coil of tube or an enclosed jacket — which contains hot water heated by the boiler. As this water passes through the exchanger it heats the water in the cylinder, but, most importantly, it doesn't mix with it. The advantage of this is that it cuts down boiler scale as only a relatively small amount of water recycles round this 'primary circuit'. At the same time it cuts down boiler wear.

An indirect hot water supply

An indirect hot water system is really two hot water systems rolled into one. It incorporates a primary circuit which runs from the boiler to a heat exchanger in the cylinder. This heats the water that is drawn off for use in the bathroom, kitchen and possibly a cloakroom – the secondary circuit.

Because it is a separate circuit, the primary system needs its own feed and expansion cistern to ensure it never runs short of water. If it did, the boiler could overheat dangerously. Fortunately, it only requires a relatively small amount of water which is being continually recycled. This reduces the effects of hard water scale and allows corrosion inhibitors to be added to prolong the life of the boiler.

Providing the boiler has the capacity, a 'wet' central heating system can be run from the primary circuit as well.

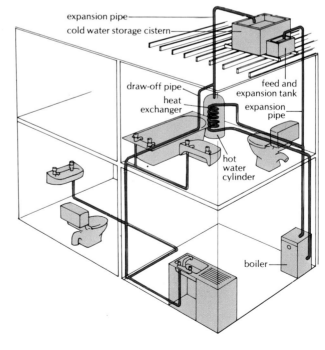

expansion pipe
cold water storage cistern
draw-off pipe
heat exchanger
feed and expansion tank
expansion pipe
hot water cylinder
boiler

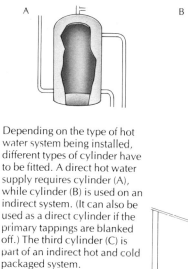

Depending on the type of hot water system being installed, different types of cylinder have to be fitted. A direct hot water supply requires cylinder (A), while cylinder (B) is used on an indirect system. (It can also be used as a direct cylinder if the primary tappings are blanked off.) The third cylinder (C) is part of an indirect hot and cold packaged system.

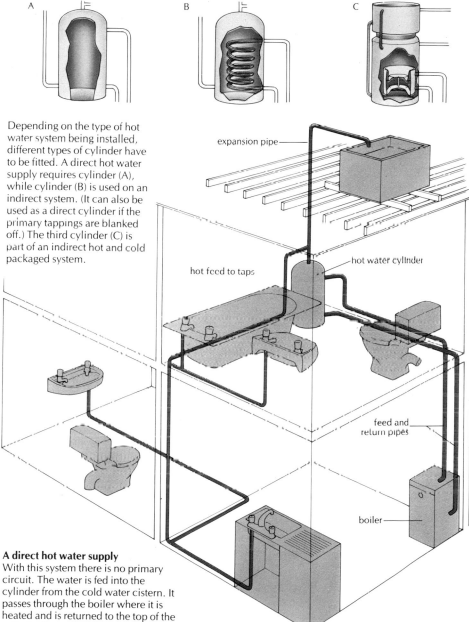

expansion pipe

hot feed to taps

hot water cylinder

feed and return pipes

boiler

A direct hot water supply
With this system there is no primary circuit. The water is fed into the cylinder from the cold water cistern. It passes through the boiler where it is heated and is returned to the top of the cylinder from where it is drawn off.

12 DRAINAGE SYSTEMS

When you pull the plug in a bath or basin, or flush a WC, it is obvious that the water has to go somewhere. It is the job of the waste water and soil system to carry this unwanted water away efficiently and hygienically through special seals called 'traps' and branch waste pipes, which feed into a large pipe known as the 'soil stack'. This is linked by an underground pipe to the main drainage system. Alternatively, in rural areas, the waste can be taken to a cess pit or septic tank. Sometimes, the system also has to deal with rainwater.

The regulations regarding the construction and adaptation of waste systems are quite stringent and are an important factor to take into account when modifying or extending the plumbing in your house. In fact, you are responsible for the waste system up to the point where it enters the main sewer; that means you have to clear any blockages. If your house was built before 1937 and has a shared drainage system where several houses discharge into the same branch of the main sewer, then the local authority is responsible for keeping the branch clean. You and your neighbours, however, have to deal with blockages and maintenance. So if you are in doubt about who is responsible for what, check with the local authority.

Single stack drainage

Now the preferred method of dealing with domestic waste, the 'single stack' system consists of a single soil stack, set inside on new buildings to give added

Single stack drainage

On a single stack drainage system, as shown on the right, there is only one soil pipe and this is sited in the house unless the system is a conversion from a two-pipe layout. Usually, all the waste pipes are fed into this pipe, except where it is more convenient for ground floor appliances to discharge over yard gullies, or for a WC to be connected directly to the branch drain via an inspection chamber.

Two-pipe drainage

The older two-pipe drainage system is illustrated on the facing page. Here, upstairs WCs discharge into a soil pipe while waste from other appliances is fed to hopper heads or trapped yard gullies.

In both types of system the soil pipe is continued above eaves level and vented to the atmosphere. It is topped by a wire mesh or plastic grille to prevent birds nesting in its end. Rainwater downpipes discharge into trapped gullies.

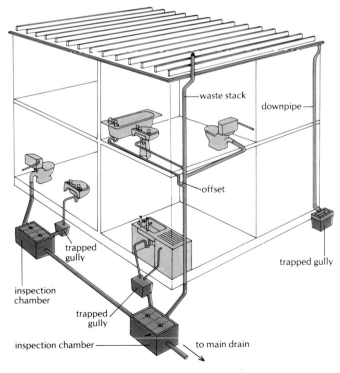

waste stack

downpipe

offset

trapped gully

trapped gully

inspection chamber

trapped gully

inspection chamber

to main drain

protection against frost. All the branch waste pipes from the bathroom, and sometimes the kitchen, discharge into this main stack. Kitchen waste water, however, can still be discharged over a yard gully, and a downstairs WC can be linked directly to the underground drainage system.

Two-pipe system

Before the single stack arrangement, the 'two-pipe' system was used. It differentiates between 'soil waste' from a WC and waste from a bath, basin or sink. Soil waste is fed into a soil pipe, which is connected directly to the underground drain, while the waste water is taken to a second pipe, which discharges over a gully incorporating a 'trap': this is an airtight seal of water in a U-shaped pipe. You'll recognise such a system by the open 'hopper head' at the top of this second pipe.

Where does rainwater go?

Rainwater falling on the roof of your home drains into the gutters and from here it is taken by a downpipe to discharge into an open gully, incorporating a trap. Then there are a number of options. Rarely is it fed into the main sewer because in periods of high rainfall it is doubtful whether that could cope with the excess water. Instead, there may be a storm drain which runs along with the main foul sewer, but if there isn't one the water is taken to a 'soakaway' (basically, a pit filled with rubble and covered with topsoil) where it can percolate back to the earth.

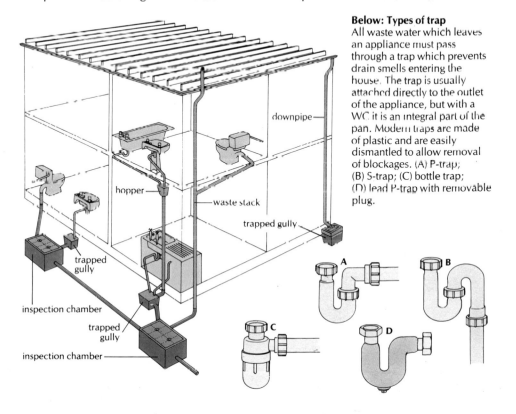

Below: Types of trap
All waste water which leaves an appliance must pass through a trap which prevents drain smells entering the house. The trap is usually attached directly to the outlet of the appliance, but with a WC it is an integral part of the pan. Modern traps are made of plastic and are easily dismantled to allow removal of blockages. (A) P-trap; (B) S-trap; (C) bottle trap; (D) lead P-trap with removable plug.

downpipe

hopper

waste stack

trapped gully

trapped gully

inspection chamber

trapped gully

inspection chamber

14 TOOLS AND FITTINGS FOR PLUMBING WORK

One of the beauties of modern plumbing is that you only have to learn a few techniques before you can tackle a wide range of plumbing jobs around the home, and you may already have many of the tools you will need. Of the few specialised tools required, most can be hired, but if you intend to carry out a considerable amount of plumbing it's worth buying at least some of them.

At the simplest level of plumbing, all the work may entail is disconnecting a couple of swivel tap connectors, unscrewing a trap, fixing a new basin where the old one used to be and then connecting all the fittings to it. Providing you can use a spanner, the job

should not pose too many problems.

Invariably, however, the work is going to be more complicated than this, and you may end up having to modify or reroute the pipe runs themselves. More ambitiously, if you want to install, say, a completely new pipe run leading to a new basin, you'll have to fit a branch to the existing system and connect up a new waste run. It is in connection with this that most of the skills — measuring, cutting, joining, bending and connecting pipework to appliances — have to be used.

Although the use of hot and cold plastic supply pipes is becoming increasingly popular and adaptors are

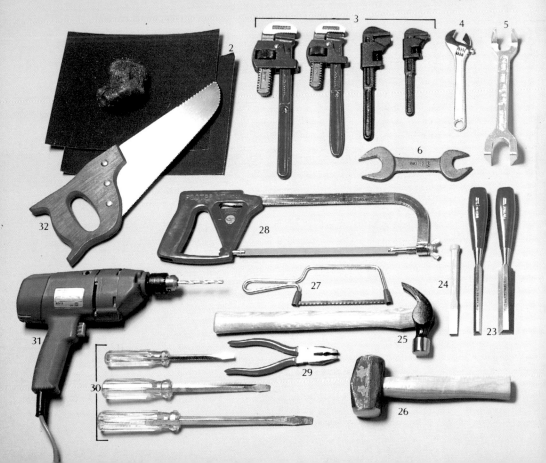

TOOLS AND FITTINGS FOR PLUMBING WORK 15

available so they can readily be connected into copper systems; because most plumbing systems are now run in copper pipe — it is easy to work with and all plumber's merchants stock it — this is what you'll be using for most of your plumbing jobs.

Copper pipe can be joined in one of two basic ways. 'Compression' joints consist of a brass or gunmetal central body with a 'capnut' at each end or outlet, which, when tightened, compresses a brass or copper ring called an 'olive' against the pipe end to form a watertight seal. Avoid the type termed as 'manipulative', which requires the pipe ends to be specially shaped to

The Plumber's Toolkit
1 Fine-grade steel wool. 2 Abrasive paper. 3 Adjustable wrenches (Stilsons). 4 Adjustable spanner. 5 Basin wrench (crowsfoot spanner). 6 Open-ended spanner. 7 Second-cut files – flat, round, half-round and tapered. 8 Spirit level. 9 PTFE sealing tape. 10 Degreasing fluid and solvent-weld cement. 11 Heat-resistant mittens. 12 Flux and brush applicator. 13 Flame-proof sheet. 14 Solder. 15 Blowlamp with gas canister. 16 Internal pipe bending spring. 17 External pipe bending spring. 18 Pipe bending machine with 15mm and 22mm formers. 19 Extendable steel tape. 20 Swarf brushes. 21 Torch. 22 Wheel tube cutter. 23 Wood chisels. 24 Cold chisel. 25 Claw hammer. 26 Club hammer. 27 Junior hacksaw. 28 Hacksaw. 29 Pliers. 30 Screwdrivers. 31 Electric drill. 32 Floorboard saw.

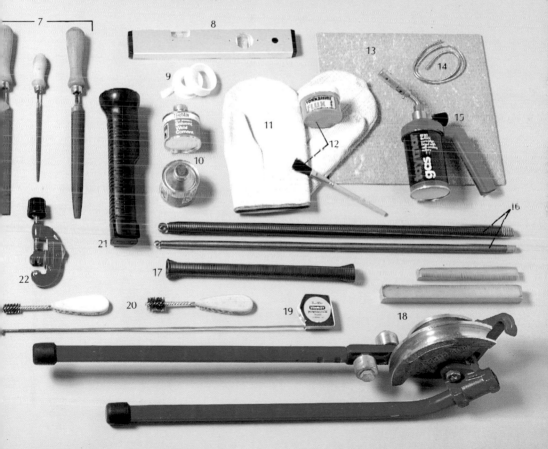

16 PLANNING PIPE RUNS

accept an internal cone fitting.

'Capillary' joints are neater looking in exposed pipe runs. They consist of a copper sleeve with outlets to take the pipe ends. 'Yorkshire' fittings have an integral ring of solder inside the sleeve which, when melted with a blowtorch, spreads around the pipe end by capillary action and forms a seal. With 'end feed' types of capillary fitting, the solder is applied separately and drawn into the space between sleeve and pipe end by capillary action. The minimum of solder should be used with these so that it doesn't flow inside the pipe. Wash out the pipes as soon as possible after making a soldered joint to remove all traces of flux.

There is a variety of fittings to enable you to join lengths of pipe: commonly straight couplers (end-to-end fixings); elbows (right-angle curves); and tees (branching into a length of pipe). Variations of these are also made for use in special situations, such as connecting pipes of differing diameters.

You will have to use plastic pipe, however, for the waste runs. If you've got a very modern system you may even find plastic hot and cold supply pipes or polythene cold water supply pipes. Again, the techniques involved in dealing with these materials are quite straightforward; if anything, they are even easier than working with copper.

Planning a pipe run

Probably the most disruptive part of any plumbing job comes when you've got to install a new pipe run. Floorboards have to be raised and exposed pipework boxed in or set in channels in the wall. Consequently, it's important to plan the route carefully and to work out exactly how you are going to proceed *before* you start. Normally, it's best to begin at the end, at the site of the new appliance, and then work towards a convenient point on the existing system where you can connect the branch. This will normally be on the pipes leading away from the cold water storage cistern and the hot water cylinder. If you work in this order, you'll only have to turn off the water supply to the rest of the house for a short time while you make the final connections between the new pipes and the old ones.

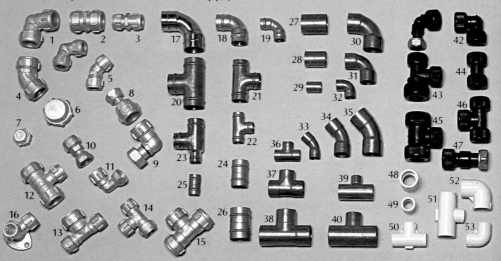

Try to plan the run so that there are as few bends as possible. A lot of right-angle bends, for example, will affect the smooth flow of the water inside and could give you a problem with air locks. Furthermore, if you use special fittings (see photographs) to make the bends then you increase the number of potential sites where leaks may occur. Once you are happy with the route, measure the length of pipe you are going to need, remembering to add extra for bends. Then check the existing pipework as it may affect the type of pipe you use to make up the new branch.

What pipework to use

There's a good chance that the main system will be copper pipe, in which case copper for the branch pipe is the obvious choice. But there is one factor you have to take into account: if the plumbing was installed before 1970, it is likely to be in imperial sized pipe — ½in, ¾in and 1in, for example, as measured across the internal diameter. Since 1970, copper pipe has been manufactured to a metric specification, but the new sizes are not direct equiva-

Common Plumbing Fittings
Compression fittings: **1** 90° elbows. **2&3** Straight couplings. **4&5** 135° bends. **6&7** Blanking-off plugs (stop ends). **8-11** Tap connectors. **12** $22 \times 22 \times 15mm$ reducing tee. **13** $15 \times 15 \times 22mm$ reducing tee. **14&15** Equal tees. **16** Elbow with backplate.
Capillary fittings (integral solder rings): **17-19** 90° bends. **20-22** Equal tees. **23** $22 \times 15 \times 22mm$ reducing tee. **24-26** Straight couplings.
Capillary fittings (end feed): **27-29** Straight couplings. **30-32** 90° bends. **33-35** 135° bends. **36-38** Equal tees. **39** $22 \times 22 \times 15mm$ reducing tee. **40** $28 \times 28 \times 22mm$ reducing tee.
Polybutylene plastic push-fit fittings: **41** Bent tap connector. **42** 90° Elbow. **43** $22 \times 15 \times 15mm$ reducing tee. **44** Straight connector. **45&46** Equal tees. **47** Straight tap connector.
CPVC plastic solvent-weld fittings. **48** $28 \times 22mm$ reducer. **49** $22 \times 15mm$ reducer. **50** Equal tees. **51** $22 \times 22 \times 15mm$ reducing tee. **52&53** 90° Elbows.
Solvent-weld waste systems: **54&55** Swept tees. **56&57** Straight couplings. **58&59** 90° bends. Push-fit (ring seal) waste systems: **60&64** straight couplings. **61** Tee. **62&63** 90° bends. **65** Swept tee.
Types of pipe: **66** Pliable corrugated copper pipe, one end plain, the other with a tap connector. **67** 15mm and 22mm copper pipe. **68** CPVC plastic pipe. **69** Polybutylene plastic pipe. **70&71** Two types of plastic waste pipe. Various types of pipe clip are also illustrated.

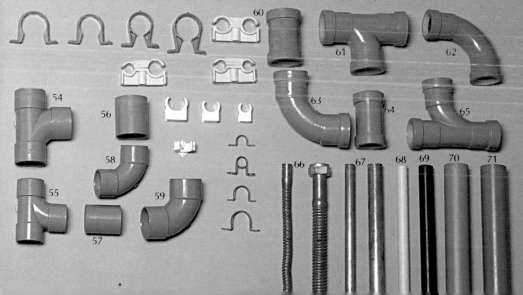

18 WORKING WITH PIPES

lents of the old because they refer to the outside diameter of the pipe. What this means, from your point of view, is that when you are connecting into ½in or 1in copper pipe you can use 15mm and 28mm compression tees, but with ¾in pipe you have to use a 22mm compression tee fitted with larger olives where it connects to the imperial pipe. With capillary fittings you can use straight adaptors to convert a short length of imperial pipe to metric and then connect metric fittings into this section.

This may all sound complicated, particularly if you have difficulty in deciding whether you have metric or imperial pipe. Therefore, because you'll be connecting mainly into 15mm (½in) or 22mm (¾in) pipe, use compression tees. They will save you a lot of time and aggravation.

If you still have dull-grey lead or clanking iron pipes, you'll know instantly that you're dealing with an old plumbing system that is nearing the end of its life. So if you are going to the trouble of installing a new branch supply, you should also consider replacing the entire system. It may sound like drastic action, but what is worse — taking this preventative measure or cleaning up after a burst pipe?

However, you may have to install,

1 Cutting copper pipe. Mark the length needed and hacksaw down the edge of tape wrapped round the pipe for a square cut.

2 Use a half-round file to remove the burr from the inside edge of the pipe and clean out any swarf from inside.

3 Remove the burr from the outside edge, then file a slight bevel so the pipe will slip smoothly into the fitting.

4 Rotate a wheel tube cutter round the pipe, gradually tightening the jaws to give a clean, square cut.

5 Insert the attached reamer into the end of the pipe and rotate it to remove any burr on the inside edge.

6 Burnish the end of the pipe with steel wool so it's ready to accept a capillary or compression fitting.

say, a washbasin, in which case you'll have to lay the branch as a temporary measure. If you are connecting into lead, you can use copper pipe for the branch, although stainless steel and polythene are also suitable. First, you'll have to set a tee into a short run of copper pipe and then fit this length into the lead supply pipe using a type of connection known as a 'wiped sol-dered' joint. There's an art to making these joints, and special tools are re-quired, so it may be best to call in a plumber to complete this part of the job for you. Likewise, connecting into iron pipe can cause problems as threads have to be cut on the ends of the iron

pipe. Again, it is wise to call in a plumber. What you must also remem-ber with iron pipe is that you can't connect copper pipe to it because it sets up an electrolytic action that will eventually cause the iron pipe to fail. Alternatively, you can use stainless steel or polythene pipes.

Connecting into polythene pipe is virtually the same as connecting into copper. Polythene pipe isn't yet pro-duced in metric sizes, so you will have to ask for ½in and ¾in (which refers to the inside diameter) instead. The com-pression tee you use to make the connection is similar to the one used for copper except that it incorporates

1 Bending copper pipe. Smear petroleum jelly on the spring, fit it in the pipe, and centre it on the bend's apex.

2 Pull the pipe round your knee, moving it slightly to prevent too tight a bend. Overbend a little, then return to the correct angle.

3 Remove the spring. If it sticks, insert a metal bar through the eye and rotate it clockwise to free the spring.

4 With a bending machine, set the pipe on the correct former under the pipe stop. Place the back guide on top.

5 Draw the two levers together so the roller works the pipe round the former until it reaches the correct angle.

6 Pliable corrugated pipe can be bent simply by hand with little effort. Don't overwork it or it will split.

20 WORKING WITH PIPES

1 Making a compression joint.
Slip the capnut and olive over
one pipe end, then offer up the
body of the fitting.

2 Mark the capnut and body
as a guide to the number of
turns you give the capnut when
tightening the joint.

3 Hold the body with a wrench
while rotating the capnut about
1½ turns with a second. Then
make up the other side.

special larger olives. The fitting also
has to be used in conjunction with
metal liners that are pushed into the
open ends of the pipes to prevent them
collapsing when the capnuts on the
fitting are tightened. The liners are
supplied with the fitting.

With some types of plastic pipe you
have to 'stick' the tee and sections of
the branch in place using solvent-weld
cement — in fact, this melts the contact
surfaces so that they bind together as
they dry. Consequently, you have to
work fairly quickly when making the

joins, but it is probably the easiest way
of getting a watertight seal. The same
technique is used to join plastic soil
pipes together.

Another type of plastic supply pipe
uses pushfit connections. These look
like plastic compression fittings, but
instead of an olive they contain an

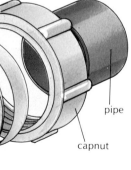

olive

capnut

pipe

pipe

capnut

olive

Making a compression joint
As the capnut is screwed on to
the body of the fitting it
compresses a metal ring — an
olive — against the end of the
socket and the pipe so forming
a watertight seal. Although
more expensive than a capillary
joint, it is much easier to use. It
may be made of brass or
gunmetal.

WORKING WITH PIPES 21

Capillary joints
The pipe must fit tightly into the socket of the fitting to ensure a good soldered seal which will be watertight.

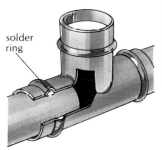

solder ring

On heating, the solder flows round the pipe and bonds the contact surfaces on cooling.

1 Making a capillary joint
Apply flux to the prepared ends of the pipes and inside the sockets of the fitting.

2 Slip the fitting over one of the pipe ends, rotating it to get an even spread of flux. Then offer up the other section of pipe.

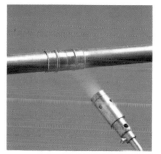

3 If the fitting has integral solder rings, play the blowlamp on the pipe and fitting until solder appears at the fitting's lip.

4 With end feed fittings, heat the pipe and fitting first. Then work solder round the joins until no more is drawn in.

O-ring seal, which forms the watertight connection. The joint is prevented from being pulled apart by a ring grip that bites into the plastic pipes.

Assembling the run
You should now be in a position to decide on the type of pipe you are going to use, how much you need and where you are going to lay it. Next comes the installation.

If you are taking the pipework under the floor, you'll obviously have to lift a few floorboards. Where the run is to be parallel to the supporting joists, clip the pipework to the side of one of them about 50mm (2in) from the top edge, so

they are well out of the way of stray floorboard nails. To avoid raising too many floorboards, you should be able to feed lengths of pipe underneath some of the boards and then lift boards at each end to give access for making any connections. Plug the end of the pipe with a rag before you slide it into position so that it doesn't scoop up any debris. Make sure, too, that you don't inadvertently damage any electricity cables.

Where the run crosses the joists this should be at right-angles. You can notch the pipes into the tops of the joists below the centres of the floorboards or drill holes for them through

22 **WORKING WITH PIPES**

1 Joining into a pipe run. Find the pipe stop positions with a dowel and mark them on the casing. Transfer the distance between them to the pipe.

2 Cut the first section of branch pipe to length. Prepare the ends and slip over the capnuts and olives before offering it up into position.

3 Turn off the supply, cut out the section of pipe and prepare the ends. There should be enough play on the run to spring the tee into place.

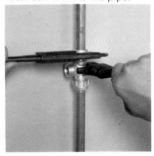

4 Use an adjustable wrench to hold the body of the tee secure while you tighten the capnuts to make a watertight seal.

5 Connect one end of the first section of branch pipe to the branch itself. Then manoeuvre the other end into the tee.

6 Make sure the branch is pressed tightly against the pipe stop. Screw on the capnut and tighten with a wrench.

the joists about 50mm (2in) from the top so they are clear of fixing nails. Any notches should be 12mm (½in) wider and 6mm (¼in) deeper than the pipe diameter to allow for movement and expansion of the pipework. Holes, likewise should be slightly larger than the pipe diameter. On the ground floor, though, if you have a sprung timber floor, you may be able to clip the pipe to the underside of the joists. This is preferable to notching or drilling since this can weaken the joists. As a well-ventilated ground floor is likely to be cold you'd be wise to insulate the pipework in this position with a proprietary lagging material. Similarly,

insulate any pipes that run through the roof space.

Taking pipework across solid concrete floors causes more of a problem. You could channel the pipe into the surface and take it round the perimeter of the room but this is a laborious and messy job. A much simpler solution is to run the pipes on the surface at skirting level where they can be boxed in unobtrusively. You face similar problems when taking a pipe run up a wall. On a brick or block wall you could channel or 'chase' the run, but this will damage any decorations, so it is not really worth considering unless you plan to redecorate. You may also

WORKING WITH PIPES 23

1 Polybutylene push-fit joints.
Cut the pipe with a hacksaw
and insert the stainless steel
sleeves into the ends.

2 Apply a bead of silicone
lubricant round the pipe end
and also coat the O-ring seal in
the socket of the fitting itself.

3 Push the pipe end firmly into
the body of the coupling. It is
prevented from coming out by
a grab ring inside the socket.

find that heat from a buried hot water
pipe might cause the plaster to crack.
The answer, again, is to run the pipe on
the surface, ideally in a corner where it
will be least conspicuous.

Avoid long, unsupported pipe runs
as these may cause knocking as the
water flows through them. Clip the
pipe at regular intervals — 1.2m (4ft)
for 15mm pipe and 1.5m (5ft) for
22mm pipe. Too high a mains pressure
may also cause knocking, and this can
sometimes be prevented by partially
closing the mains stopvalve.

Working with pipe

Regardless of the material the pipework
is made of, the first thing you'll have to
do with it is to cut it to the required
lengths. Accurate measuring is important, as is cutting squarely. A good
hacksaw with a sharp blade will handle
most types of pipe, but with copper
you'll find a wheel tube cutter a good
investment, particularly if you have a
lot of cutting to do.

Joining and connecting Joining lengths
of pipe is a straightforward task. Use
compression fittings for copper, stainless steel and polythene. You can also
use capillary fittings for copper and
stainless steel. When you are working
with certain types of plastic pipes (such

as UPVC and CPVC), you'll need to
make solvent-weld joins. Pushfit fittings are used on plastic waste pipes,
and for one particular type of plastic
hot and cold supply pipes. You can
also use conventional compression fittings for the latter, but in addition you
need to insert a metal liner into the
mouth of the pipe. Most connections to
waste traps and to appliances are made
with screw fittings.

Bending pipe No pipe run is perfectly
straight; sooner or later it will have to
change course. Lengths of copper pipe
can be bent using a bending spring —
the internal type that slips inside the
pipe is now the most common — or a
pipe bending machine, but you can
also use special angled couplings.
Polythene pipes can be bent by hand
but they need to be firmly clipped in
position to prevent them springing
back. With stainless steel, it's best to
use the special fittings.

Draining the pipes

Before you install any new branch
pipe, you will have to drain that section
of the supply pipe you intend to cut
into. Most connections will be made
on the supply pipes that leave the cold
water storage cistern and the hot water
cylinder. Don't forget that you must not

24 WORKING WITH PIPES

1 Solvent-welding waste pipes
Cut the waste pipe to length
with a hacksaw, using a paper
template to get a square end.

2 Use a file to remove any
rough edges, then chamfer the
ends so the pipe can be slotted
easily into the fitting.

3 Rub the ends of the pipes and
the insides of the sockets on the
fitting with steel wool to give a
key to the contact surfaces.

4 Wipe degreasing fluid over
the contact surfaces with a
cloth so that they are
thoroughly cleaned.

5 Use the brush applicator in
the can to apply solvent-weld
cement liberally to the pipe and
the inside of the sockets.

6 Push the pipe into the socket,
twisting it to spread the cement.
Clean off surplus cement and
allow it to set.

connect into the cold pipe that feeds
the hot cylinder or the primary circuit
that runs between the boiler and the
hot cylinder.

With luck, there should be gatevalves on the supply pipes leading
away from the cold cistern. Just close
these off and open the hot and cold
taps in the bathroom and the hot tap
over the kitchen sink. The hot water
pipes will only flow for a very short
while because the draw-off point is at
the top of the cylinder, and with the
cold feed turned off there is little
pressure in the system to generate the
flow. Consequently, you won't empty
the cylinder of water. If you have to
drain the cylinder, there should be a

draincock (tap) near the bottom so you
can run a hose from this to an outside
gully.

Once the water in the pipes has
stopped flowing, you can insert the
branch, but be prepared with some old
rags to catch the inevitable small trickle
of water as the pipe is cut. When you
restore the supply, make sure all the
joins are watertight before concealing
the pipework.

If there are no gatevalves, you'll have
to drain the cold water cistern. First of
all, you've got to cut off its supply.
Rather than turn off all your water by
closing the main stopvalve, it's best to
tie the cistern's ballvalve to a length of
wood spanning the tank so it cannot

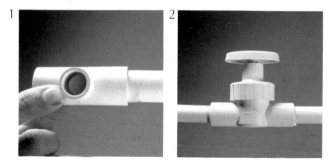

Solvent welds in CPVC pipe
Basically this is the same operation as that used for waste pipes. Chamfering is unnecessary, nor do you have to remove surplus cement. **(1).** Allow this to set for 1 hour before running cold water and 4 hours before running hot. You can fit gate valves **(2)** provided they are fitted with stubs of CPVC pipe.

open when the taps in the house are turned on. By only draining down through the cold taps, you won't waste expensively-heated hot water.

You may also want to connect into the rising main above the stopvalve. This won't cause any problems if there is a draincock directly above the valve. If there isn't, then the job is a little more complicated. Turn off the supply as normal, and then release the top nut making the compression fitting between the valve and pipe leading to the cold water cistern. You should be able to spring the pipe from the body of the valve and quickly put a funnel connected to a length of hose under it. Lead the hose to a bucket nearby to catch the water as it drains out. Remake the compression joint when the branch is complete; then you can restore the supply. But before doing this, why not take the opportunity to fit a draincock — the type with a straight compression coupling — into the run to make draining easier in future.

Push-fit joints in waste pipe
This process is slightly more involved than making similar connections in plastic supply pipe, because this pipe expands when heated by hot water and this has to be allowed for.

When making a join, first clean all the contact surfaces with special cleaner. Chamfer the pipe ends and apply a coating of petroleum jelly **(1)**. Check to see that the sealing ring is correctly positioned in the socket, then push the pipe in firmly so that it butts up to the pipe stop **(2)**. Make sure the pipe is positioned squarely before marking the position of the socket mouth on it **(3)**. Now withdraw the pipe so that the pencil mark is about 10mm (3/8in) from the socket **(4)**. This should provide sufficient room for the pipe to expand.

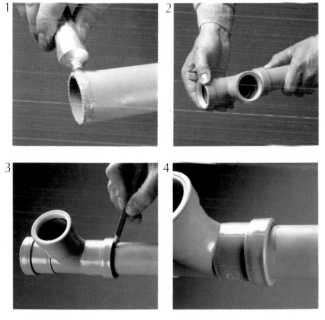

26 THE KITCHEN SINK

The kitchen taps and sink are probably the most used part of a domestic plumbing system. The sink, particularly, takes a battering, being used to wash everything from dishes and clothes to paint brushes. The waste system, too, has to deal with all kinds of debris that it would be best off without.

It is remarkable how well a sink — whether enamelled pressed steel or stainless steel — will stand up to this treatment. Eventually, however, the scratches will show, the enamel will chip and hard water scale will build up. The net result is a rather unhygienic-looking sink. The answer is to fit a new one.

You may even have an old glazed earthenware sink with bib taps projecting from the wall instead of being fixed to the sink itself. While these sinks have their charm, they are not really practical as they have no attached drainer. If

Types of inset sink top

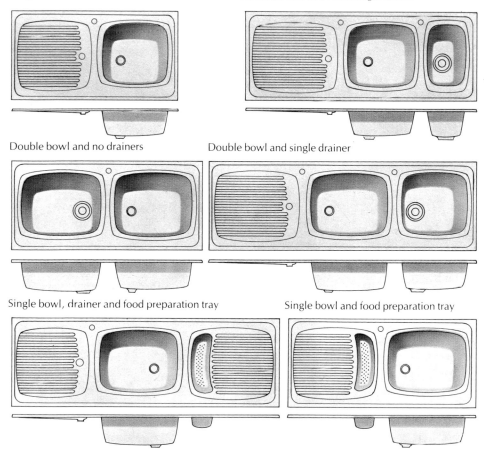

Single bowl and drainer

Bowl and half bowl with single drainer

Double bowl and no drainers

Double bowl and single drainer

Single bowl, drainer and food preparation tray

Single bowl and food preparation tray

How a mixer tap works

The flow of hot and cold water into the body of the mixer is controlled by individual taps. But the supplies are not actually mixed here. Instead, they are conducted along separate channels in the spout and only come together as they leave its nozzle. The reason for this is to prevent hot water from being siphoned into the rising main.

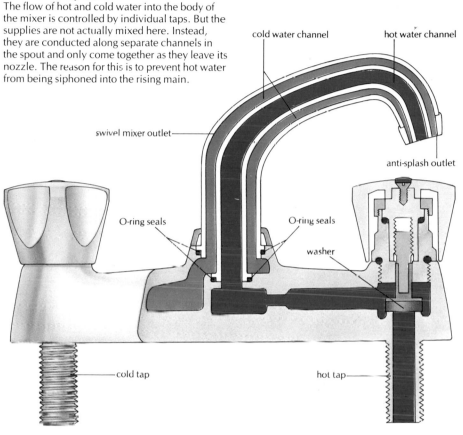

cold water channel

hot water channel

swivel mixer outlet

anti-splash outlet

O-ring seals

O-ring seals

washer

cold tap

hot tap

you are modernising your kitchen, such a sink is something you'll almost certainly want to replace.

There is a wide variety of sink tops from which to choose. These range from a traditional one-piece sink and drainer that fits over the top of a kitchen unit to an inset sink with double bowls, or just inset bowls where the surrounding worktop acts as the drainer. Stainless steel is still popular, but modern chip-resistant enamel and plastic sinks are becoming common and are available in a choice of colours. Likewise, ceramic sinks are making a comeback.

New taps?

If you're replacing an old sink, the chances are that you'll also need new taps; and if you're replacing bib taps with taps fixed to the sink, you'll also have to modify the plumbing to take the pipework under the sink. Taps vary in price enormously, but basically you have the choice between individual pillar taps and mixers with a swivel spout — the most practical type to get. The most modern are known as 'monobloc' taps and they often incorporate a useful rinsing brush to remove suds from dishes.

28 INSTALLING A NEW SINK

The first job is to remove the existing sink top, but don't do anything until you've turned off the hot and cold supplies to it and drained the pipes (see page 23); then you can undo the connections. With the old sink out of the way, you'll be able to see if the hot and cold supply pipes and the waste outlet match the new sink. If they don't, you'll have to adjust their positions. This could lead to some awkward pipe bending, so as an alternative try using pliable copper pipe to bridge the gap between the supply pipes and the tap inlets. This may mean cutting the supply pipes back a little so you can get a smooth run on the bendable pipe.

Fortunately, much of the installation work can be done before the sink is put in position — which makes the job easier and quicker. You then have to set the top on its unit or over the hole in the worktop if it's an inset sink. When it's fixed in place, you can reconnect the supply pipes and screw on the trap and waste. The trap should be a 'P', bottle or dip-partition type (see page 13). All that remains then is to restore the water supply and check for leaks.

If you previously had a glazed earthenware sink, it will probably have had a lead trap and waste pipe. You will have to replace this in its entirety using a 38mm (1½in) diameter trap and pipe, which is usually made of UPVC. As the waste will discharge into a gully, make sure that the end of the new pipe is carried below the grille but stops short of the water level. You can cut the grille to admit the pipe or simply buy a replacement grille with a suitable hole in it.

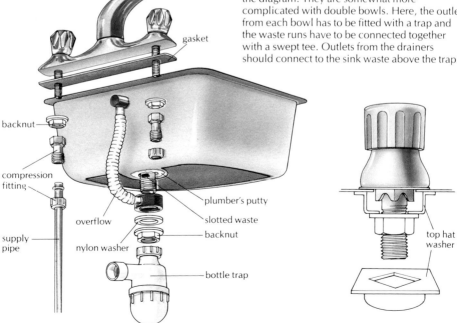

The plumbing connections for a single bowl sink are relatively straightforward and are shown in the diagram. They are somewhat more complicated with double bowls. Here, the outlet from each bowl has to be fitted with a trap and the waste runs have to be connected together with a swept tee. Outlets from the drainers should connect to the sink waste above the trap.

mixer

gasket

backnut

compression fitting

supply pipe

overflow

nylon washer

plumber's putty

slotted waste

backnut

bottle trap

top hat washer

INSTALLING A NEW SINK 29

1 Wind PTFE tape round the waste outlet thread and apply plumber's putty beneath the flange, unless a plastic or rubber washer is supplied.

2 Insert the waste into the sink's outlet. Smear jointing compound on the washer and slip it over the outlet thread, compound side down.

3 Align the outlet in the collar of the overflow hose with the slot in the waste. Then fit and secure the backnut to hold everything in place.

4 Attach the other end of the overflow hose to the outlet at the top of the bowl by screwing it to a small grate and chain stay, attached to the plug.

5 Next, position the mixer tap and slip top hat washers over the inlet tails. Screw on the backnuts, then connect lengths of pliable pipe.

6 Set the inset sink on a rubber gasket or clear mastic bed when positioning it over the hole in the worktop. Clip it so it cannot move.

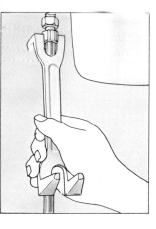

7 Bend the pliable pipes so that they will meet the supply pipes. Screw on the bottle trap and connect the waste pipe.

A crowsfoot spanner, also known as a basin wrench, is little more than an open-ended spanner with the ends bent at 90°. This means that it doesn't have to be held at right angles to a nut in order to turn it. Consequently, it is ideal for loosening those seemingly inaccessible connections at the back of a bath or basin when you want to remove the taps. If backnuts are difficult to shift, apply extra leverage to the free end by slotting in a spanner and putting pressure on that.

30 WASTE DISPOSAL UNITS

A waste disposal unit is a boon to the modern, efficient kitchen because it disposes of kitchen waste — which usually ends up in the dustbin — hygienically and quickly. You simply tip the waste down the sink, turn on the cold tap and the unit, and steel grinders reduce the matter to a slurry that's safe to flush into the waste system.

Obviously, the unit can't deal with all kitchen waste. Large bones, cartons and fibrous matter which will bind round the grinding blades still have to be thrown in the bin.

A waste disposal unit must be connected to the waste outlet of the kitchen sink before the trap, and to accommodate the unit the sink needs a larger outlet hole than normal — one with a diameter of 89mm (3½in) instead of 38mm (1½in). However, there is a model available which does connect to the normal-sized outlet.

If you are replacing a sink, you can buy a replacement with the correct sized opening, but if you've got a stainless steel sink which you are keeping, you can cut a larger hole using a special tool. It's not possible to cut a larger hole in a ceramic sink or an enamelled sink, so replace both.

When you've connected the unit to

1 Remove the old waste from the sink and check that the diameter of the outlet is big enough to take the new sink bush of the disposal unit.

2 The waste disposal unit is attached to the sink bush via a clamp seal and suspension plate. These are held together by a circlip.

3 First, apply a layer of water-resistant sealant under the flange of the sink bush, then set the bush in the sink outlet, bedding on the sealant.

4 Slip the seals and plates over the bush, fit the circlip and slightly tighten the grub screws on the suspension plate.

5 Fit the seal over the outlet bend and locate the bend in the top housing. Secure it by screwing on the outlet plate.

6 Set the flat seal in the recess in the top of the housing and smear it with a silicone lubricant.

Right: Waste disposal units consist of three sections: a clamping system which holds the unit securely round the sink waste outlet; a top housing where the waste is ground to a slurry; and a sealed motor unit to which the grinding blades are connected.

the new outlet, screw on the trap. Use an ordinary 'U' or 'S' trap rather than a bottle trap, as the latter is more prone to blockages. The waste pipe should slope at about 15°, and if it discharges into an open gully it's important the pipe ends below the grille so there is no chance of the slurry creating a blockage. Otherwise, the waste pipe can be connected to a single stack system.

The unit also needs an electric power supply. You can run this as a spur from a ring circuit to a switched fused connection unit with a neon indicator. The flex from the disposer has to be wired to this. Never undertake any electrical work unless you're sure you know what you're doing. Follow the maker's instructions, and consult a qualified electrician regarding any cross-bonding required with the pipework.

Site the connection unit in an accessible place but away from the sink unit. On some disposers you have to fit a magnetic cap into the sink outlet before the unit will operate.

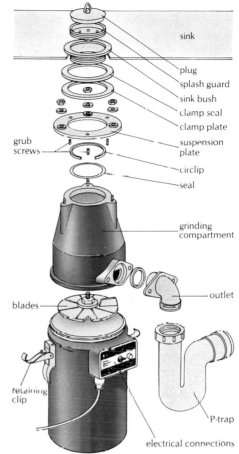

sink

plug
splash guard
sink bush
clamp seal
clamp plate
suspension plate
circlip
seal

grub screws

grinding compartment

outlet

blades

P-trap

retaining clip

electrical connections

7 Offer up the top housing so that the studs locate in the suspension plate grommets. Then bolt the unit to the plate.

8 Swivel the housing to align the outlet pipe with the trap. Tighten the grub screws to prevent further rotation.

9 Fit an O-ring seal to the motor housing and clip it to the top housing, making sure that the switch panel is accessible.

32 WASHING MACHINES AND DISHWASHERS

Although you can still buy twin tub washing machines, front- and top-loading automatics that include a spinner are now considered a necessity in many households. The convenience of having a washing machine permanently connected to your home's plumbing system, ready for use at the touch of a button, is obvious and many people are also realising the benefits of an automatic dishwasher.

Because they are usually stationed in the kitchen or utility room where there are hot and cold supply pipes and a waste system to hand, plumbing in these appliances is a job you can easily tackle yourself.

The plumbing connections

Washing machines and dishwashers are plumbed in similarly. Most need a supply of hot and cold water, although some just need a cold supply, which they heat internally to the required temperature. Ideally, the pressures of the hot and cold supply should be the same. However, when putting a machine in the kitchen it's usual practice to take the cold water from the supply to the kitchen tap, which is under mains pressure, and the hot water from the hot supply to the kitchen hot tap, which is under lower gravity pressure. Most machines can cater for the pressure difference, but check the manufacturer's instructions.

When you are teeing off the hot and cold supplies, you're almost certain to need 15 × 15 × 15mm compression or capillary fittings to start the branch. You'll also need to install 'mini-stop-valves' on the individual runs, which will enable you to screw the flexible rubber hoses of the appliance directly to their outlets. When the machine needs to be serviced, all you have to do is to turn off the valves and disconnect the hoses.

The waste outlet hose of the machine simply hooks into the end of a 'stand-pipe'. This is a 600mm (2ft) long vertical pipe with an internal diameter of 35mm (1⅜in). It should have a 75mm (3in) 'deep seal' trap at its bottom end, which is then connected to the drainage system. Sometimes, the trap comes as part of the standpipe itself. This arrangement prevents water being siphoned out of the machine while it's in operation.

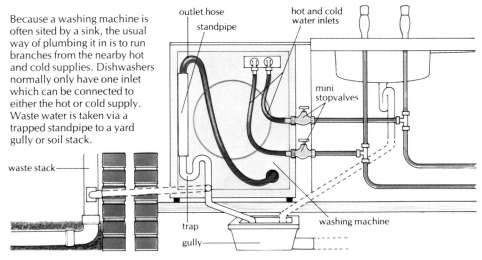

Because a washing machine is often sited by a sink, the usual way of plumbing it in is to run branches from the nearby hot and cold supplies. Dishwashers normally only have one inlet which can be connected to either the hot or cold supply. Waste water is taken via a trapped standpipe to a yard gully or soil stack.

outlet hose
standpipe
hot and cold water inlets
mini stopvalves
waste stack
trap
gully
washing machine

WASHING MACHINES AND DISHWASHERS 33

1 Knock a hole in the outside wall and feed through the waste pipe from the standpipe position inside. Then fit a 90° bend to the end.

2 Direct the waste pipe into a nearby gully so that it stops below the level of the grille, which you will have to cut to fit round the pipe.

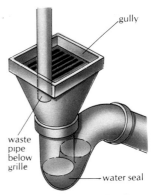

The waste pipe should end above the water.

3 Make up the standpipe and connect the trap to the waste run. Bracket the pipe to the wall and check the top is 600mm (2ft) above floor level

4 Run 15mm branch pipes to the site of the appliance from suitable hot and cold supplies. The pipes should stop just to one side of the machine.

5 Fit washing machine or mini-stopvalves to the branch pipes. Some have a backplate for wall fixing. Make sure they are easily accessible.

6 Fit the machine's inlet hoses directly to the outlets of the stopvalves. (Note: dishwashers only have a hot or cold fill.)

7 Feed the outlet hose 150mm (6in) into the standpipe. The hose must never be immersed when the appliance empties.

8 Adjust the feet of the appliance so that it's level. Then plug it into a 13 amp socket and test the system.

34 DEALING WITH HARD WATER

When you look at a glass of water it's difficult to believe that this seemingly inoffensive liquid could actually be attacking your plumbing system. But if you live in a hard water area — and about two-thirds of homes in Britain have such a supply — this may be happening.

The visible signs of hard water are all too apparent: scale builds up round the element in an electric kettle; soap is difficult to lather; and baths, sinks and basins have to be cleaned constantly to remove the tide marks that remain when water drains away. So just think what is happening on the inside of pipes and boilers as scale begins to accumulate. The plumbing system has to work harder to achieve the same results as the flow of water is gradually reduced. Sooner or later, something is bound to fail, and often it's the element of an immersion heater that burns out fighting to overcome the blanket of fur.

What is hard water?

Hardness in water is due to the presence of certain magnesium and calcium mineral salts. However, there are two types of hard water: temporary and permanent. If temporary hard water is heated to above 60°C (140°F) scale is deposited and the water starts to lose its hardness. It is this form of hard water that can play havoc with a hot water system. This doesn't happen with permanent hard water — heating it doesn't cause the fur deposits that block the pipes, yet the water still retains its other hard qualities.

Dealing with the problem

There are several things you can do to alleviate the effects of hard water. First, you can set the boiler or immersion heater thermostat to 60°C (140°F) — in soft water areas it can be 10°C (18°F) higher — and, if you haven't already got one, install an indirect hot water

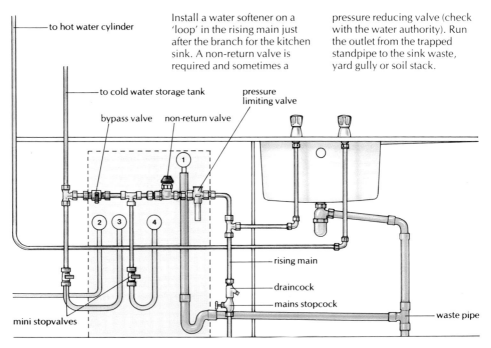

to hot water cylinder

Install a water softener on a 'loop' in the rising main just after the branch for the kitchen sink. A non-return valve is required and sometimes a

pressure reducing valve (check with the water authority). Run the outlet from the trapped standpipe to the sink waste, yard gully or soil stack.

to cold water storage tank

pressure limiting valve

bypass valve   non-return valve

mini stopvalves

rising main

draincock

mains stopcock

waste pipe

DEALING WITH HARD WATER 35

 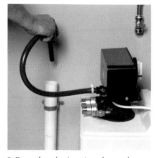 

1 Connect the flow and return pipes into the rising main together with any bypass valve, non-return valve and pressure reducing valve supplied.

2 Run the drain pipe from the unit to a trapped standpipe, which in turn can either discharge into a yard gully or soil pipe via a strap boss.

3 Run the overflow pipe sloping downwards to a visible point on an outside wall where you will be able to see any water flowing from it.

4 Next, cut the flow and return hoses to length and fit connectors to the ends. Do not forget to incorporate their sealing washers.

5 Connect the other ends of the hoses to mini-stopvalves on the branch pipes. Fit a filter washer on the 'flow' to stop debris entering the unit.

6 Wire the softener directly to a switched fused connection unit fitted with a 3 amp fuse, or connect it to a socket via a plug with a 3 amp fuse.

cylinder (see page 61). Scale inhibitors can then be added to the water in the primary circuit. You can also put water softeners in your washing water; indeed, many cleaning agents already contain them.

However, the most effective way of dealing with hard water is to install a chemical mains water softener, which will take all the hardness out of your supply. It does this by converting the offending magnesium and calcium salts into sodium salts, which do not form scale. Normally, it is plumbed into the rising main after the branch to the cold

tap over the kitchen sink, as it's thought healthier to drink hard water in preference to soft. The unit also needs an electric power supply and a connection to a drain. Given these requirements, probably the easiest place to install it is in the kitchen, but it doesn't have to be there; you can install it anywhere convenient.

Water softeners are expensive and some people see them as something of a novelty. However, they may reduce household bills — for soap and the like — and so pay for themselves in about six to seven years.

36 FITTING A NEW HANDBASIN

The other major area in your home where water is required is the bathroom, and often this may include a WC. Plumbing jobs in the bathroom range from the simple — dealing with a leaking tap — to the more complicated tasks of installing a new shower or a completely new bathroom suite. In the latter case, if you are putting the new bath, basin or WC in the same place as the old one, you are unlikely to breach any building regulations. However, if you want to alter the position of any of the appliances you should consult your local building inspector, who will be able to advise you on whether your plans meet the various regulations. Mostly, he'll be interested in how you intend to run the waste pipes and connect into the drainage system.

Replacing the handbasin

One of the most common jobs in the bathroom is to replace the handbasin, which may just be old fashioned or, worse, cracked.

Obviously, the first job is going to be the removal of the old basin. Once you've turned off the water supplies, use a special 'crowsfoot' spanner to undo the tap connectors, then unscrew the waste. This will leave the basin free to be unscrewed from the wall and lifted clear of any mounting brackets. As with a kitchen sink, you may now have to alter the position of the supply pipes and waste to accommodate the new basin. If you are lucky, everything will match up but don't forget that the shanks of new taps are shorter than those of old ones. Fortunately, you can buy specially extended tap connectors to get round this problem.

The step-by-step photographs show you how to go about fitting a new basin. As you can see, much of the work — fitting the taps and outlet — can be done on the floor. If you are

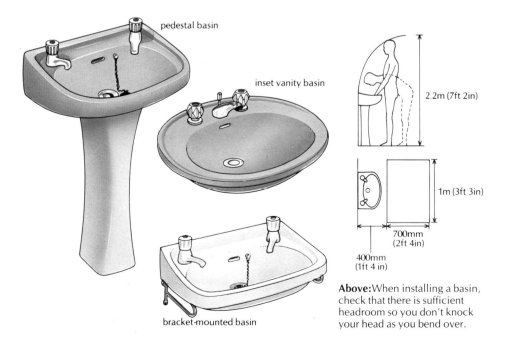

pedestal basin

inset vanity basin

2.2m (7ft 2in)

1m (3ft 3in)

700mm (2ft 4in)

400mm (1ft 4in)

bracket-mounted basin

Above: When installing a basin, check that there is sufficient headroom so you don't knock your head as you bend over.

FITTING A NEW HANDBASIN 37

1 Mark the position of the special support studs on the wall. Drill and plug the fixing holes, then screw the studs in place with a wrench.

2 Before mounting the basin, attach the taps. Bed them on gaskets or plumber's putty, then tighten up the backnuts to hold them securely in place.

3 Unscrew the heads of the studs, allowing you to slip the basin over the shanks. Check for level, then secure with the nuts set on the studs.

4 Insert the waste, bedded on plumber's putty or any gasket supplied. Sandwich the angle bracket between washers and loosely screw on the backnut.

5 Fix the bracket to the wall, then tighten the backnut fully. At this stage, you can also connect up the hot and cold water supply pipes.

6 Finally, install the new waste run, taking it to a suitable outlet point such as a gully or soil stack. At the basin fit a bottle trap to the waste outlet.

installing a basin in a new location, there are a few other points you've got to bear in mind.

First, you've got to decide how you are going to dispose of the waste water. The waste run to the soil stack or downpipe shouldn't be longer than 2.3m (7ft 6in) and should contain as few bends as possible; and on single stack drainage systems the gradient of the branch should be between 1° and 2½°. If the waste run is less than 1.7m (5ft 6in) long, you can use 32mm (1¼in) waste pipe; anything longer than this should be run in 38mm

(1½in) diameter pipe. In both cases, these must run from a 75mm (3in) deep seal P-trap. On basins it's common to fit a bottle trap.

The final task is to connect the waste run into the drainage system. If the basin is upstairs and you've got two-pipe drainage, the waste can discharge into a hopper head. With single stack drainage you'll have to break into the plastic soil pipe using a special 'strap boss' (see page 56). If the basin is on the ground floor, the waste can either discharge over a yard gully or be linked to the soil stack.

38 **REPLACING A BATH**

Despite the convenience of showers, many people still like to relax in a bath. Yet if you install a bath/shower mixer tap, instantly you have the advantages of both. Lately, there have been some new developments that have aroused more interest in baths. Not least of these is the spa bath, in which a 'whirlpool' (created by streams of air and water fed into the bath through a series of inlet ports) certainly provides a bath with a difference.

If you have an old cast iron bath, you may want to change it because the enamel has started to wear away — a problem that occurs particularly below the spouts of taps and round the waste outlet. Probably the main reason for changing the bath, however, is because you want to modernise the bathroom completely.

If you're doing this, there's the opportunity to change the position of some or all of the fittings. A word of warning here: always bear in mind the space you need to use the bath, basin and WC, and how you are going to run the plumbing.

To help solve the planning problem, one manufacturer produces a special bathroom planning kit. It incorporates a grid on which you can position scale outlines of bathroom fittings, allowing you to double check that everything you want will fit in the room with space to use it. You can also plan the hot and cold supply runs and the route for the waste pipes. The irony is that you may find the original layout was, after all, the most logical for the space available and so end up simply replacing everything in situ.

The techniques of replacing a bath are much the same as those for a basin, except that you're dealing with a larger fitting and all the plumbing is likely to be less accessible. The best plan, therefore, is to complete as much of the

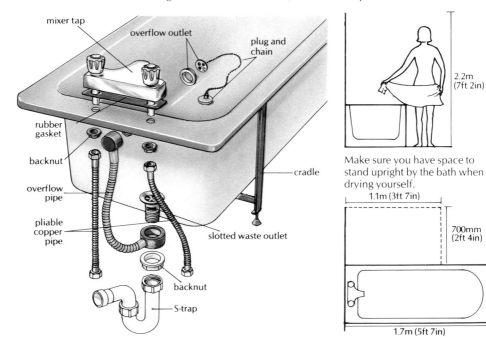

mixer tap
overflow outlet
plug and chain
rubber gasket
backnut
overflow pipe
pliable copper pipe
cradle
slotted waste outlet
backnut
S-trap

2.2m (7ft 2in)

Make sure you have space to stand upright by the bath when drying yourself.

1.1m (3ft 7in)

700mm (2ft 4in)

1.7m (5ft 7in)

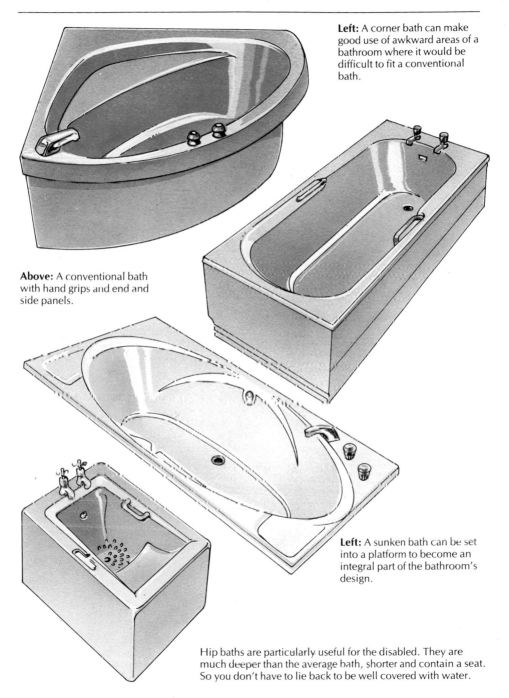

Left: A corner bath can make good use of awkward areas of a bathroom where it would be difficult to fit a conventional bath.

Above: A conventional bath with hand grips and end and side panels.

Left: A sunken bath can be set into a platform to become an integral part of the bathroom's design.

Hip baths are particularly useful for the disabled. They are much deeper than the average bath, shorter and contain a seat. So you don't have to lie back to be well covered with water.

40 **REPLACING A BATH**

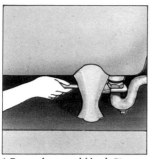

1 Removing an old bath First, turn off the water supply, then deal with the lead trap. Unscrew the nut holding the trap to the bath with an adjustable spanner.

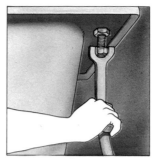

2 Use a crowsfoot spanner to get to the tap connectors, but there should be space to use an adjustable spanner on other couplings.

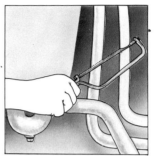

3 If the nuts won't move, you'll have to cut the pipe, yet this won't matter if you have to re-route the runs to the new bath.

work as you can on the new bath before you have to turn off the water and take out the old one.

Preliminary work

As you'll probably be fitting new taps to the new bath, this is one of the first jobs you can do. Individual pillar taps or a mixer normally sit on rubber gaskets. You have to slip a deep 'top-hat' washer over the shank beneath the tap to bridge the unthreaded section. Then, the backnut can be wound tight to hold the tap firmly in place. With individual taps, grip each spout while you are turning the nut so the tap isn't tightened up crooked. At this stage, it is also worthwhile connecting lengths of hand-bendable corrugated pipe to the tap tails, making it much easier to connect to the supply pipes later on — rigid lengths of pipe can be difficult to connect up if there's even a slight misalignment.

Next, you can fit the bath outlet. Its rim, or 'flange', should be bedded down on a rubber washer, so place this round the underside of the flange. Push the outlet into place and screw up the backnut. The final job is to connect the overflow to the bath and link a special collar sometimes called a 'banjo unit',

on the other end of the flexible plastic overflow pipe, to the waste outlet. Align the hole in the collar with the slot in the waste outlet.

Removing the old bath

Disconnecting the old bath from the plumbing may prove something of a problem, particularly if it's been there for a good many years; the tap connectors may be impossible to free even with a crowsfoot spanner. Some penetrating oil may do the trick, but if not, the simplest solution is to cut through the pipes. There should be enough room to use a mini-hacksaw, but if there is not, use a file saw. Don't be too worried about making the cuts square; you can tidy up the ends later.

Check to see if there is a separate overflow and if there is, disconnect it. As the overflow on a modern bath is connected to the waste outlet just above the trap, you can remove this pipe and block up the hole through the wall. Remove any screws attaching the bath to the floor and walls.

You should be able to remove the old bath and dispose of it now, but if it's cast iron you're going to need help — it will be very heavy. The alternative is to break it up where it stands so you

REPLACING A BATH 41

1 First mount the bath in its cradle, then fit the mixer. Bed it on a gasket, slip the tails through the fixing holes and secure the backnuts.

2 Fit the slotted waste to the bath outlet. Slip on the overflow collar and screw on the backnut. Connect the other end to the bath overflow.

3 Fit pliable copper pipe to the mixer inlet tails so that connections to the hot and cold supplies will be easier when the bath is positioned.

4 Position the bath and adjust the cradle's feet to get it level. Screw through the feet and bracket to the wall.

5 Join the pliable pipes to the supply pipes using compression couplings. This avoids any scorching of the bath.

6 Finally, screw on the P-trap to the bath's waste outlet. Link the trap to the waste pipe with a ring-seal, push-fit connection.

can carry it out piece by piece. You'll need a club or sledge hammer to do this, but you must cover the bath with an old carpet or blanket to prevent shrapnel flying in all directions.

With the old bath out of the way, there will be plenty of space to make any alterations to the supply pipes, but if you are using flexible copper pipe you may not have to do this. You might just have to cut back a small section of supply pipe so you can get a smooth bend on the flexible pipe. Now is also a good time to install a new plastic waste run in 38mm (1½in) diameter pipe.

When you are happy that all the pipes are in the right place, you can install your new bath. Check that it's level, evenly supported by its cradle and that there is sufficient clearance underneath for the P-trap, which should have a 75mm (3in) seal.

Join the supply pipes to the flexible pipes using compression fittings; if they are the old imperial ¾in pipes all you have to do is fit larger olives to the fittings. It may be a bit awkward wielding spanners in the restricted space available, but if you were to use a blowlamp to make a solder coupling you would risk damaging the bath, particularly if it is made of plastic.

42 FITTING A NEW WC

Browsing through the glossy brochures of any manufacturer of bathroomware will immediately bring home to you just how far bathroom design has advanced in recent years. Although the shapes of baths and basins have been refined, if not dramatically altered, it is in the design of WCs that the changes are most apparent. Nowadays, one of the main reasons for fitting a new WC suite is because the existing one is conspicuously old-fashioned, bulky and noisy when it's flushed. The old type of high-level cistern with its pull chain is ungainly when compared with a modern low-level, streamlined design operated by a neat handle or plunger.

Replacing a WC is not as difficult a job as at first it may appear, particularly if you intend to connect the pan to the same waste outlet. In fact, one of the most common jobs is to convert a high-level set-up into a low-level one. You needn't change the pan to do this, so simplifying matters even further. All you'll need is a slimline cistern known as a 'flush panel', which will only stand out 115mm (4½in) from the back wall. You need this type of cistern because with a high-level arrangement the pan is set close to the wall; if you installed an ordinary cistern low down you wouldn't be able to raise the seat fully.

After you've turned off the water supply and drained the cistern, disconnect the pipework attached to it. You may find it quicker to cut through it if the nuts on the connectors are encrusted in paint and difficult to turn. The cistern can then be removed.

Next, fix the flush panel in place — the top should be at about waist height. When you are happy with its position, you can install the internal flushing mechanism and the operating handle. The ballvalve can also be attached at this stage. All that remains then is to connect up the new pipework. The 15mm cold water supply could be an

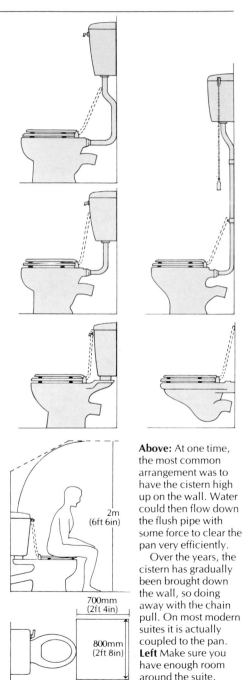

Above: At one time, the most common arrangement was to have the cistern high up on the wall. Water could then flow down the flush pipe with some force to clear the pan very efficiently.

Over the years, the cistern has gradually been brought down the wall, so doing away with the chain pull. On most modern suites it is actually coupled to the pan. **Left** Make sure you have enough room around the suite.

2m (6ft 6in)

700mm (2ft 4in)

800mm (2ft 8in)

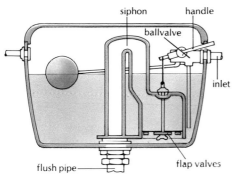

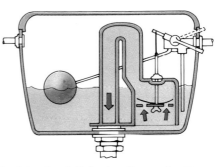

On a modern piston-type cistern, when the lever is pulled down, the plunger in the cylinder is drawn upwards to lift water over the siphon bend. Siphonic action continues the flow, with water passing through the flaps in the plunger until the cistern empties.

Different types of WC pan

Right: In a conventional flush-down pan, it is the force of water entering from the cistern that enables the pan to be cleared of waste and the trap replenished with clean water. The water flows in two streams around the rim of the pan, meeting at the front, efficient operation depending on correct cistern capacity and length and diameter of the flush pipe.

WASHDOWN WC

Below: In a single-trap siphonic pan, it is the gravity pressure of water that starts the flow through the trap. But the outlet is designed to create a siphonic action which further helps to draw the water through.

SINGLE-TRAP SIPHONIC WC

Below: With a double-trap siphonic pan, when water flows into the pan, air is drawn out of the chamber between the traps through a valve. This, combined with the gravity flow of water, sets up a strong siphonic action to clear the trap.

DOUBLE-TRAP SIPHONIC WC

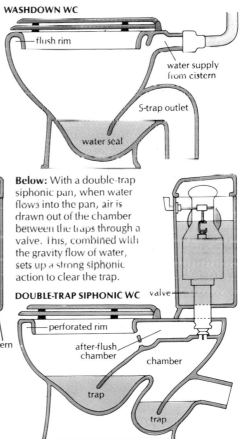

44 FITTING A NEW WC

extension of the feed to the old cistern. As this may provide you with a somewhat conspicuous and ugly pipe run, however, you may choose to branch off the cold water network in a more convenient and less noticeable place. Now is also a good time to install a stopvalve on the pipe run just before it enters the cistern to connect to the ballvalve. This will enable you to isolate the cistern to make repairs without the need to turn off all the cold water supply.

Finally, you'll have to link the pan and cistern with a new plastic feed pipe and drill a hole through the outside wall to take the overflow pipe.

Installing a close-coupled WC

If you're going to install a completely new close-coupled suite, you will have to deal with the pan as well as the cistern. Start by removing the cistern (as previously described), then turn your attention to the pan, which you can bail out now or wait until later.

First of all, you've got to break the seal between the integral trap of the pan and the waste outlet. The pan may have a P-trap, in which case the waste pipe will probably go straight back through the outside wall — the usual arrangement with an upstairs WC — or

it may have an S-trap, with the waste pipe rising up through the floor. After you've broken the seal with a club hammer and bolster chisel, check to see if the pan is screwed to the floor. If it is, remove the screws and you should then be able to lift clear the pan and the feed pipe from the old cistern.

When you've cleaned and prepared the rim of the waste outlet, you can stand the new suite in place and mark the fixing holes for the cistern and pan. These can be drilled and plugged (for wall and solid floor fixing) when the suite is taken away. The easiest way of connecting the trap to the collar of the waste pipe is to use a flexible plastic connector. This can be fitted before the pan is finally screwed to the floor.

Next, you can slot the cistern into the coupling at the back of the pan and fit the flush mechanism and ballvalve. All that remains is for the 15mm cold water supply pipe to be connected up, for a new hole to be drilled for the overflow, and for the overflow pipe itself to be laid in.

Changing the position

If you want to change the position of the WC, it's important that you don't contravene the Building Regulations. So it's a good idea to contact your local

1 Turn off the supply to the cistern, then flush to empty it. Disconnect the pipes, cutting them free if necessary.

2 Remove any fixing screws driven through the back of the cistern so you can lift the tank carefully off the brackets.

3 Unscrew and take out the pan. Block up the outlet socket on the soil pipe while you drill and chisel out the old mortar.

building inspector to check that your plans are in order. There are a number of points that you ought to consider. The most important of these are concerned with the waste run, which should always be in 100mm (4in) UPVC plastic pipe.

Most WCs are situated on an outside wall and you won't run into too many problems if you just want to move the suite along that wall. You can extend the waste along the skirting from where it enters the room and then use a 90° connector to take it into the back of the pan. The waste pipe branch should not be longer than 6m (20ft), should in-corporate a 100mm (4in) fall and should contain as few bends as possible. Usually, the pipe can be neatly boxed in to conceal it.

If you want to install a completely new WC and waste branch, you'll have to make a new connection into the drainage system either at the soil stack or, if the WC is on the ground floor, at an inspection chamber. This is a some-what ambitious job and one where it is worth taking professional advice — particularly if you have a two-pipe drainage system with a cast iron soil pipe. See page 56 for more information about work involving soil stacks.

1 Here, the cistern and soil pipe will be hidden behind a false wall. Join the pan to the soil pipe with a 'Multikwik' connector.

2 Mark the position of the cistern within the framework. It needs to be screwed to the back wall as well as being supported on brackets.

3 Connect the overflow pipe to the top of the cistern, then run it at a slight downward gradient to a place where it will be visible on an outside wall.

4 Set the siphon over the outlet in the base of the cistern, then assemble the linkage for the flushing mechanism.

5 Run in the 15mm cold water feed pipe attaching it to the ballvalve assembly. Make sure the capnut is tightened fully.

6 Connect the flush pipe to the bottom of the siphon and the pan. Test the system, adjusting the float arm if necessary.

46 INSTALLING A BIDET

Until recently, the bidet was regarded as something of a curiosity in the bathroom. But it is, in fact, a useful and versatile appliance.

There are two types of bidet to choose from. The over-rim or wash-basin bidet operates, as its name implies, like an ordinary basin. In fact, it is plumbed in identically, often sharing the hot and cold supplies to the basin. This bidet can have individual pillar taps or a mixer, and the most up-to-date version of the latter has a spray at the end of the nozzle.

The second type of bidet is known as a rim supply bidet with ascending spray, and is somewhat more difficult to install than the basin type. When you turn on the taps, warm water is first directed round the underside of the rim, so making it more comfortable to sit on. The supply can then be switched to a spray in the bottom of the bidet, which works like a fountain. This means that the rose is covered in water

when the pan is filled. So in order to reduce the risk of soiled water being siphoned back into the plumbing system, the hot water supply must be taken directly from the hot cylinder and the cold from the cold cistern. This is where the extra plumbing work comes in, and how much you have to do depends on where the tanks are situated. The base of the cold water cistern should also be 2.75m (9ft) above the inlet of the bidet.

When you come to deal with the waste, you'll need to fit a 75mm (3in) deep seal trap if you've got single stack drainage, and the 32mm (1¼in) waste pipe — 38mm (1½in) if the run is between 1.7m (5ft 6in) and 2.3m (7ft 6in) maximum — must connect directly into the stack. For information about this, see page 56. On a two-pipe drainage system, you need only fit a 50mm (2in) shallow trap, and the waste can discharge directly into a gully or hopper head.

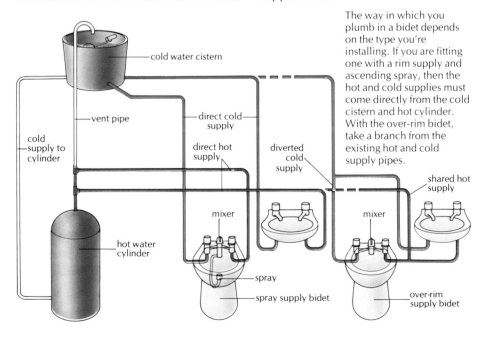

The way in which you plumb in a bidet depends on the type you're installing. If you are fitting one with a rim supply and ascending spray, then the hot and cold supplies must come directly from the cold cistern and hot cylinder. With the over-rim bidet, take a branch from the existing hot and cold supply pipes.

cold water cistern

vent pipe

direct cold supply

cold supply to cylinder

direct hot supply

diverted cold supply

shared hot supply

mixer

mixer

hot water cylinder

spray

spray supply bidet

over-rim supply bidet

INSTALLING A BIDET 47

1 Link the mixer unit, incorporating the pop-up waste control, to the tap headgear and screw on the flange to hold it firmly in place.

2 Screw on the nuts to hold the taps in position, making sure there are washers on both sides of the bidet, then fit the decorative covers.

3 Work a layer of plumber's putty round the flanges of the slotted waste outlet and spray rose and set them in the base of the bowl.

4 Slip a washer over the waste outlet and screw on the waste extension piece. On some types you have to fit a backnut before doing this.

5 Next, connect the bendable metal hose to the bottom of the rose and take the other end to the spray outlet on the mixer control unit.

6 Link the control rods of the pop-up waste. Connect one end to the plunger and insert the other into the waste outlet, where the plug is fitted.

7 Make up a P-trap and screw it to the waste outlet. You may need to offset it slightly to avoid the waste mechanism.

8 As with a bath, you will find it easier to connect the water supply to the bidet if you fit pliable pipe to the tap tails.

9 Fit the waste pipe to the trap. Connect the hot and cold supplies and finally screw the pan to the floor.

48 INSTALLING A SHOWER

To many people, there are few plea-
sures that come near to relaxing in a
hot bath; indeed, in many homes a
bath is still the only means of providing
a complete wash, but the increasing
popularity of the shower is steadily
changing that.

Baths do have a number of draw-
backs and it is these that shower
manufacturers have been able to capi-
talise on. Baths take up a considerable
amount of space; they take time to fill;
and they use a lot of expensively
heated hot water.

Showers, on the other hand, are
more hygienic, they are easier to control, and quick to use. The other main
benefit is that the average shower uses
only 20 per cent of the water a bath
would. Furthermore, because a shower
cubicle doesn't take up much room —
about 760mm (2ft 6in) square is all
that's required — it doesn't have to be
sited in the bathroom. Providing you
can meet the plumbing requirements,
you can put a shower in a bedroom, on
a landing, in a downstairs utility room
or cloakroom, or even under the stairs.
In fact, by not installing the shower in
the bathroom you could avoid the
usual bathroom traffic jams first thing
in the morning!

The simplest shower

If you don't want to go to the trouble
(and expense) of installing an indepen-
dent shower, the easiest alternative is
to make use of the bath and its plumbing. You're probably familiar with the
basic rubber hose shower attachment,
together with its limitations, but you
can buy a bath/shower mixer that
simply replaces the bath taps. This has
a spout which is used to fill the bath
and a flexible hose leading to a shower
rose, which you can hook on the wall.
A plunger mechanism lets you decide
whether you want to work the shower
or run water into the bath. As you can

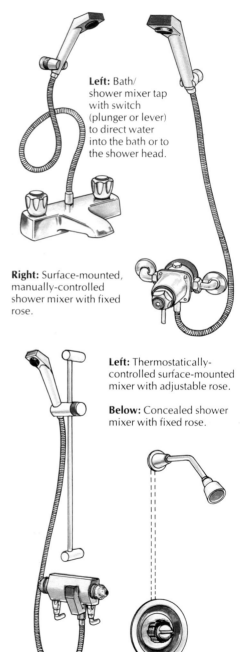

Left: Bath/
shower mixer tap
with switch
(plunger or lever)
to direct water
into the bath or to
the shower head.

Right: Surface-mounted,
manually-controlled
shower mixer with fixed
rose.

Left: Thermostatically-
controlled surface-mounted
mixer with adjustable rose.

Below: Concealed shower
mixer with fixed rose.

If you have a conventional plumbing system with a cold water cistern in the loft and a hot water cylinder, say, in an airing cupboard, feeding the hot taps, then fitting a shower mixer in a cubicle or over a bath is virtually the same as installing any tap outlet. But to minimise the problem of the water running cold or scalding hot when the bath taps are operated, take the cold supply direct from the cistern and the hot from the feed and expansion pipe above the main draw off point.

Since the supply of water to the shower will be under gravity only, you should also check that there is a pressure head of at least 1m (3ft 3in) between the bottom of the cistern and the shower rose. This will ensure a good strong flow of water to the rose when the tap is turned on.

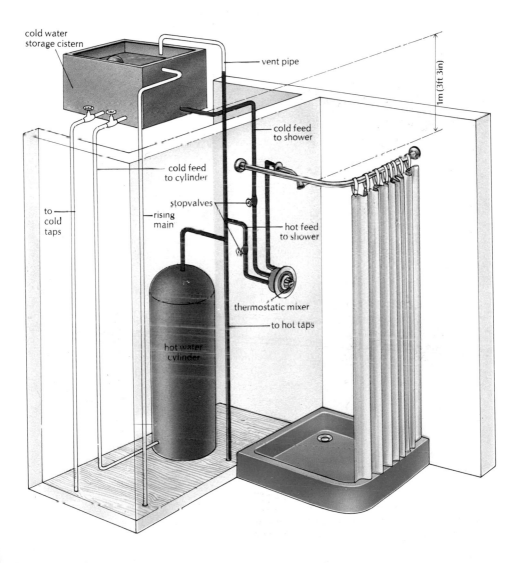

cold water storage cistern

vent pipe

1m (3ft 3in)

cold feed to shower

cold feed to cylinder

to cold taps

stopvalves

rising main

hot feed to shower

thermostatic mixer

to hot taps

hot water cylinder

50 INSTALLING A SHOWER

see, this arrangement provides an easy and cheap shower. The only extras you'll need are a shower screen or a curtain to prevent water splashing over the bathroom floor.

As with all showers, there must be sufficient pressure in the water supply in order to get a good flow of water through the rose. With the basic bath/shower mixer you will also have to adjust the taps delicately to obtain the water pressure and temperature you require for a comfortable shower.

But because of the way your bath plumbing is likely to be organised, someone may only need to flush a WC or run the hot tap over the kitchen sink to cause sudden variations in the temperature and flow. You could even scald yourself as a consequence. The simplest way of avoiding these unpleasant occurrences is to buy a thermostatic mixing valve.

An independent shower

If you want to install a shower in its own cubicle, or a separate shower over the bath, you have a choice of two methods. You can either use your home's hot and cold water system to supply the water or, if this isn't possible, you can install a new electric shower. So before you go out and buy a shower unit, you must first check out the plumbing system.

Fortunately, the design of most domestic plumbing systems is such that you should have few problems in fitting an ordinary mixer shower (see page 49). Because the cold water cistern is normally in the loft, in all probability its base will be at least 1m (3ft) above the shower rose which will give sufficient 'head' of water. The cold water supply to the mixer is taken from a new 15mm pipe, which you'll have to run from the cold water cistern. The hot water supply should be taken from the vent and draw-off (expansion) pipe that leaves the top of the hot water cylinder.

You may be tempted to take an easier alternative by branching off the bath or basin hot and cold feeds, but don't. The mixing valve works most efficiently if the water supplies have an equal and constant pressure. By connecting the pipes directly to the cold water cistern and hot water cylinder outlet, you will ensure the minimum of temperature fluctuation.

Mixer taps

Once you've got the water to the site of the new shower, you can connect the pipes to an array of different styled controls and shower roses — some

1 Fitting a shower rose. Fit the glide, which incorporates a rose holder, to the runner, then screw this bar to the wall.

2 Mark the position of the hose outlet on the wall and run the pipe from the mixer to this point behind the cubicle.

3 Connect the pipe to the hose union with a compression joint concealed by a plate. Screw the other end to the rose.

1 A surface-mounted mixer. Screw the backplate to the wall. Fit the unit, mark the pipe entry points and run them in.

2 You will find it easier to make the supply connections at the mixer if you first disconnect the chrome stubs.

3 Screw the hose on to the mixer outlet and take the other end to the rose, which can either be fixed or adjustable.

mixers can be recessed into a wall; others are surface mounted. Looks will play an important part in deciding what you choose because the unit is going to be visible, but on the practical side make sure the mixer can deliver about 4.5 litres (1 gal) of water a minute at about 43°C (109°F) — a typical good, soaking shower.

There are two types of mixer from which to choose: manual and thermostatic. There are two controls on the manual mixer, one for the rate of flow and the other to set the temperature of the water. However, the mixer will still be affected by fluctuations in the supply, so it's important that you run the

supply pipes as previously described. For ease, you could branch off the hot system close to the shower but then you've got to put up with the shower suddenly going cooler when someone turns on a hot tap. This is where the more expensive thermostatic mixer scores, because it allows you not only to control the flow but also to set the temperature of the water and maintain it at that level despite fluctuations in the supply. These mixers usually incorporate a cutout device which shuts down the shower if, for example, the cold water supply pipe bursts, leaving only very hot water flowing through the mixing valve.

1 The shower tray. Fit the waste outlet to the tray and connect a shallow-seal P-trap to it. This should have an inspection eye.

2 Temporarily set the tray in place so you can work out where the waste pipe needs to be run in relation to the trap.

3 The tray should be set level on a brick or timber plinth if it does not incorporate its own mounting frame.

52 INSTALLING A SHOWER

You may be unlucky and not have a typical plumbing system, in which case you'll have to make some modifications, install a shower pump or fit a mains pressure electric shower (see page 54 for details).

The commonest problem you'll have to deal with is a cold water cistern that isn't high enough above the shower rose to give a sufficient 'head' of water. If the cistern is already in the loft, you could raise it up on a platform. In itself, this isn't a difficult plumbing job — just fiddly. Similarly, if the cistern is above a hot water cylinder in the airing cupboard, then it is a quite straightforward job to transfer it to the loft to give the head of water you need.

However, moving the cistern into the loft is out of the question if you have a flat roof. The easiest solution, here, is to install a shower pump, and this will also involve you in a small amount of electrical work. If you are not confident that you can do the electrical work, consult a qualified electrician.

You might have a combination hot and cold water storage unit. With this, you've got another problem apart from low water pressure, because the cold water capacity of the unit is only large enough to supply the hot cylinder and not any taps or other outlet points. In fact, the other taps will probably be supplied direct from the mains. What you have to do, therefore, is to install a new small cold water storage cistern, which will be used exclusively for the shower, and a shower pump. The alternative is to install an electric shower, virtually bypassing the entire plumbing system.

If you've got a multi-point gas heater, and so no storage cylinder, you should still be able to fit a shower, but take professional advice on this because you may have to fit a 'pressure governor' on the cold water supply to the heater and shower.

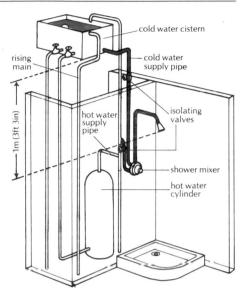

It may be necessary to raise the cistern in the loft (see above), otherwise there are alternative shower arrangements using shower pumps. These need to be wired to a ring circuit via a double-pole, switched fused connection unit.

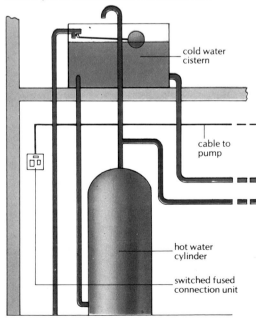

Right: On packaged hot and cold water systems, install a small cold water cistern to feed the shower and take the hot water direct from the cylinder. You'll also need to install a shower pump (see below).

Below: Depending on the type, shower pumps can be located either somewhere convenient outside the shower cubicle (on the floor under a bath, for example) or within it, protected by a waterproof casing. If the pump is operated by a flow switch, so that it comes on when the shower is turned on, it should be plumbed into the outlet side of the mixer running to the shower rose.

However, there are pumps that can be turned on and off independently of the flow by a 12 volt switch or a cord-operated switch which takes its power from the pump itself. The pump is connected to the hot and cold supplies before they reach the mixer.

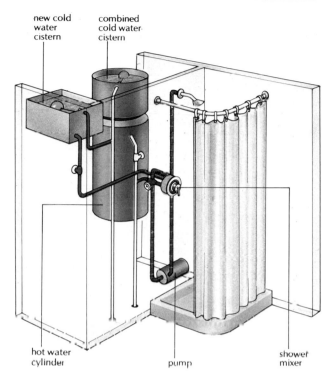

new cold water cistern

combined cold water cistern

hot water cylinder

pump

shower mixer

Pump concealed behind cubicle wall.

Pump inside cubicle in waterproof casing.

Pump outside cubicle on the floor

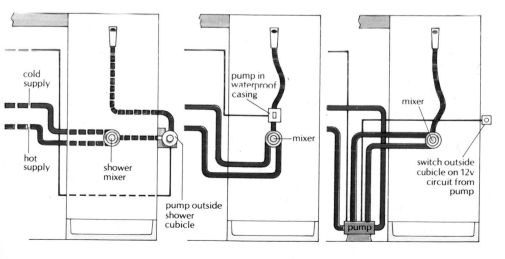

cold supply

hot supply

shower mixer

pump outside shower cubicle

pump in waterproof casing

mixer

mixer

switch outside cubicle on 12v circuit from pump

pump

54 ELECTRIC SHOWERS

If you can't run a shower easily from your existing plumbing system, it may be worth considering fitting an electric shower. This is economical to run as you only heat the water you use. All it requires is a 15mm cold water feed, branching from the rising main, and a separate 30A electrical circuit. Because all the workings are concealed behind a waterproof cover, the unit can even be installed inside a shower cubicle.

As the shower contains its own heating element, you don't have to worry about a hot water supply, but the need to heat the water instantly reduces the flow through the unit. Therefore, because it emits less water than an ordinary shower, it's considered less 'wet', but still adequate for giving a refreshing spray of water.

When you buy one of these showers, it's best to buy one with a temperature stabiliser so that the temperature of the water doesn't fluctuate if the pressure in the rising main drops. Normally, if the pressure drops too low, a pressure sensor will turn off the shower before the water has a chance to overheat and scald you.

Containing the water

Perhaps it is stating the obvious to say that a shower cubicle must be waterproof, but it is a point worth reiterating. Water has a way of seeping through the tiniest of gaps and can give rise to all sorts of problems if not checked.

If you are installing a shower over a bath, the majority of cubicle problems are solved immediately, as you're using a ready-made system to dispose of the waste water. All you need is a waterproof curtain or a plastic or safety glass screen to provide the sides of the shower cubicle.

Installing a new cubicle requires a

Types of shower cubicle

Left: A shower can be fitted in a small alcove. **Right:** Freestanding units are quickly installed. **Below:** A simple screen turns a bath into a shower cubicle.

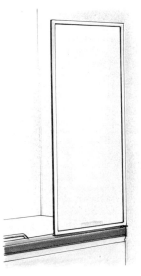

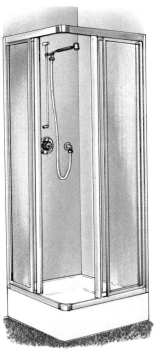

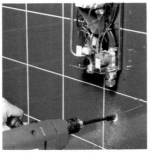

1 Mark the fixing holes for the unit and the cable entry hole. Drill these holes and mount the shower at about chest height on the cubicle wall.

2 Drill an access hole for the cold water feed. Use a flat bit on cubicles lined with laminate and a masonry bit on a ceramic tiled surface.

3 Connect up the 15mm cold feed to the inlet of the shower. Do not forget to protect tiles from a blowlamp when making capillary joints.

4 Next, screw the shower hose on to the outlet. The connector must incorporate a washer in order to make an effective watertight seal.

5 Connect the 6mm² 2 core and earth cable to the terminals of the unit. Run the other end to a 30 amp double-pole cord-operated switch.

6 Fit the rose to an adjustable glide. Then screw on the casing, checking that the fingerplates link correctly to the controls behind.

little more work, particularly in deciding how you're going to run the waste pipe to the domestic drainage system. You can get round some of the installation work by using a shower cabinet with a combined tray and sides, but in most cases you'll be installing a ceramic or plastic tray and erecting one of the vast selection of proprietary screens round it. It's vital that the join between the two is waterproof and this is usually ensured using mastic.

You could, if you wish, build your own solid surround to a shower tray, facing it with ceramic tiles or laminate.

In all cases, make sure you leave enough space under the tray to attach the 50mm (2in) shallow seal P-trap. This may mean raising the tray on bricks, wooden blocks or metal brackets. A shower tray with a corner outlet will improve access to the trap in case of blockage. The waste pipe should have a diameter of 38mm (1½in) and it should run with as shallow a fall as possible to connect into a soil stack or to discharge over a hopper (see page 56). Don't position your shower so that you have to cut through any joists in order to install the waste pipe.

56 DEALING WITH THE WASTE

When you start dealing with soil stacks, and waste systems generally, plumbing work can get more involved. This is not so much because of the techniques required — plastic waste systems, after all, have either solvent weld or pushfit joints — but because of the type of system you may have and the building regulations regarding how waste pipes should be run. Consequently, it's worth restating that before you start to modify or add to your plumbing system you should make sure you know what you are going to do with the waste.

Single-stack drainage

If you've got single-stack drainage (either a system that was installed when the house was built, or a properly carried out conversion from a two-pipe system), your problems are reduced considerably. The diagram opposite shows exactly what your system should conform to, but there are a number of points to bear in mind if you are carrying out any modifications.

If you are installing a new WC upstairs, try and connect it into the existing soil waste branch with a 'running swept tee' joint so you don't actually have to make a new connection into the stack. If this isn't possible and the waste pipe comes from the other direction, you'll have to insert a 'double equal branch' into the stack so that both runs can join at the same level. This may prove difficult if the soil pipe is inside the house and boxed in, as you may be limited for space. Even if the pipe is on the outside, you may still have problems when trying to manoeuvre the junction into place after you've cut the pipe.

The waste from a new shower or basin could also be linked to a convenient waste pipe, but with a basin you may be able to arrange the waste run so that it can connect directly to the stack. Here, you can use a fitting known as a 'saddle' or 'strap boss' (a curved plastic plate with a socket inlet in the centre).

Before you begin work on modifying or installing a new above-ground drainage system, however, you should draw up plans of the system and submit them to your local building inspector. It is essential that any work you do is with his approval and complies with local building regulations.

The same type of connection problem exists at ground floor level, but here you can link a new WC directly to the underground drain, either at an existing inspection chamber or at a new one which you'll have to build. However, this is a job which goes beyond basic domestic plumbing, and you will definitely have to consult your local authority for advice and approval.

Two-pipe drainage

Because the soil stack of a two-pipe system is likely to be made of cast iron, you should not try to make additional connections to it. Instead, organise your new waste runs so that they will connect into existing waste branches inside. There are plastic pushfit connectors to link old cast iron pipes to new plastic runs; if the old pipe is bigger than the 100mm (4in) diameter pipe now used, there's an offset connector that will make the conversion. New shower and basin wastes can discharge into the hopper at the top of a downpipe, or they can be taken to a downstairs gully.

However, if you intend making alterations in a two-pipe system, you should consult your local building inspector. Hopper heads, for example, are considered unsanitary and are no longer encouraged; it may be better, therefore, to modify the system or to convert it to a single-stack layout, depending on where the appliances are sited. In turn, this may mean replacing the traps and waste and soil pipes.

DEALING WITH THE WASTE 57

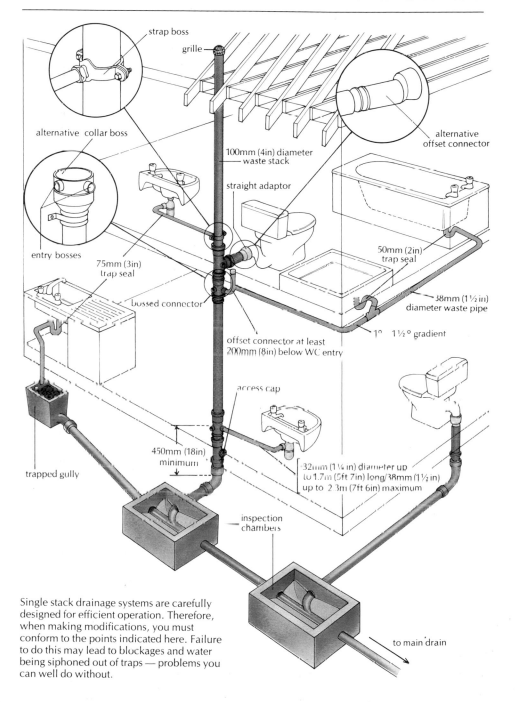

strap boss

grille

alternative collar boss

100mm (4in) diameter waste stack

straight adaptor

alternative offset connector

entry bosses

75mm (3in) trap seal

bossed connector

50mm (2in) trap seal

38mm (1½in) diameter waste pipe

1°   1½° gradient

offset connector at least 200mm (8in) below WC entry

access cap

450mm (18in) minimum

trapped gully

32mm (1¼in) diameter up to 1.7m (5ft 7in) long/38mm (1½in) up to 2.3m (7ft 6in) maximum

inspection chambers

to main drain

Single stack drainage systems are carefully designed for efficient operation. Therefore, when making modifications, you must conform to the points indicated here. Failure to do this may lead to blockages and water being siphoned out of traps — problems you can well do without.

58 COLD WATER CISTERNS

Most domestic plumbing work entails installing or repositioning fittings — a basin may need to be moved or a new shower fitted — but to carry out most of this work, you won't have to touch the cold water cistern. Usually, if you want a new supply of water, all you have to do is run a branch from a convenient point on the pipe network. Only if you are installing a shower will you need to run a new cold feed from the cistern; and then there are alternative plumbing arrangements which mean you don't necessarily have to do this.

Modern cold water cisterns are usually made from plastic, although you may well find that you've got one made from glassfibre, asbestos cement or zinc coated (galvanised) steel. The cistern is normally sited in the loft, but if your house has a flat roof it's quite likely to be over the hot water cylinder in the airing cupboard. In fact, as stated previously, there are package systems where the cold water cistern and the hot water cylinder are incorporated in the same unit.

Basic maintenance
Although you won't have much call to get to the cistern, it still requires periodic checks, say, about two or three times a year.

Plastic and asbestos cement cisterns don't corrode, but if you've got a grey galvanised steel tank then this is something worth looking for. These cisterns do have a long life, but the increasing use of copper pipe can lead to problems. Copper and zinc in the same system, with a slightly acid mains supply of water, can set up an electrolytic action which causes the zinc to dissolve. Without its protective coating the steel cistern is defenceless against the effects of corrosion.

However, you can still use copper pipes with this type of cistern if you suspend what's known as a sacrificial

anode (a piece of magnesium) in the water. As this is more prone to electrolytic action than zinc it will slowly dissolve instead, so leaving the zinc layer intact and the cistern protected. If you do find that the zinc has started to wear away, drain down the cistern, scrape off any rust and then treat the inside with a couple of coats of bituminous paint. Make sure it's the type made for use in cisterns as this won't taint the water.

The other item to check is the ballvalve. Although this may be working, it only takes a small piece of grit brought up in the rising main, or a speck of hard water scale, to stop the valve opening and closing properly. If the taps aren't being used too regularly, you may not notice that anything is wrong until you draw off a lot of water and suddenly the taps go dry because the cistern hasn't been filling properly. Repairing the ballvalve, however, is a simple job (see page 73), and all it may need is a good clean or the washer replaced. At worst, replacing the entire ballvalve is a straightforward job.

Installing a new outlet
Making a new outlet in your cistern to feed, say, a new shower shouldn't cause any real problems, although you may find working in the restricted space of a loft a bit awkward.

With plastic and galvanised cisterns, you can use an electric drill fitted with a special tank cutter accessory or a hole saw of the appropriate size to make the outlet 50mm (2in) up from the bottom. Then, install the outlet fitting and screw up the swivel connector to secure the pipe. Remember, when working with plastic cisterns, never to use jointing compound — only PTFE tape — to make watertight seals. Always use compression fittings instead of capillary types when working in a roof space, as the flame from a blowlamp could ignite

COLD WATER CISTERNS 59

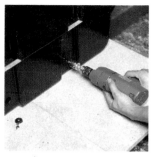

1 Set the new cistern on an 18mm (¾in) platform. Drill all the outlet holes 50mm (2in) above the base using a hole saw cutter fitted to a drill.

2 Wind PTFE tape clockwise round the thread of the tank connector. Slip on a washer and push the thread through from inside the cistern.

3 Slip another washer over the thread, then tighten the connector nut to form a watertight seal. Repeat this operation for all outlets.

4 Connect short stubs of pipe to the connectors using compression fittings. You can wrap PTFE tape round the pipe ends for a good seal.

5 Connect gatevalves to the ends of the pipe stubs before continuing the runs. These are better than stopvalves, as they help prevent hammering.

6 When cutting the inlet hole at the top of the cistern, support the flexing wall with a block of wood held against the inside face.

7 Fit a high-pressure ballvalve to the inlet as you would a tank connector. Then connect the rising main to this.

8 Similarly, drill a hole and fit a plastic tank connector to the overflow outlet. Run this pipe to discharge at the eaves.

9 Fit the lid and make a hole in it to take the expansion pipe from the hot water cylinder. Then lag the tank and pipes.

60 COLD WATER CISTERNS

the dust there and cause a major fire.

Making an outlet in an asbestos cistern may be a little more difficult because of the nature of the material and the extra thickness of the walls. However, by marking the hole on the side, at least 100mm (4in) up from the bottom, and drilling a series of small holes round the inside of the perimeter, you should be able to tap out the centre and clean the edges with a half-round or round file. You'll find drilling this material much easier if you first grind the tip of the bit to a sharp point. When you fit the connector, make sure you use soft washers against the asbestos surface to get a really watertight seal.

Installing a new cistern
Just because you've got an old cistern doesn't necessarily mean you've got to replace it, but if the existing one is badly corroded, or too small for your present needs, a new one is called for. You may be able to repair holes in a galvanised cistern with epoxy resin filler, but this is only a stopgap measure and it is best to cut your losses and replace the cistern entirely.

You will also have to install a small cistern as a feed and expansion tank for the boiler if you're converting a direct hot water system into an indirect one. If you have a combined cold water cistern and hot water cylinder unit, the capacity of the cistern is usually only great enough to serve the cylinder, so if you want to install a conventional shower, you'll have to put in another small cold water cistern to serve it.

Plastic cisterns are the best to use; and one with a capacity of 228 litres (50 gal) is sufficient for household needs. Because the plastic is flexible, such cisterns can usually be manipulated even through small trap doors to the loft. However, if access to the loft is a problem, take up two 114 litre (25 gal) cisterns and connect them together in series. The ballvalve inlet should go at the top of one of the tanks and the outlets should be taken from the bottom of the other. The cisterns are connected by a length of 28mm pipe.

You can also use this 'tandem' arrangement if, when you've pushed the old cistern to one side, there's not the space to install a new larger one. You're in luck if the old cistern can be lowered through the trap door because most cisterns were installed before the roofs that cover them.

You may find it easier to drill the holes for the pipe connections to the cistern before setting it on a chipboard or timber platform. The expansion pipe from the hot water cylinder should discharge over the rim, while the inlet pipe from the rising main should be installed as high on the side as possible. The overflow pipe should be fitted 25mm (1in) below the outlet point of the ballvalve, which should be the high-pressure type. The supply pipes must be connected 50mm (2in) up from the bottom of the cistern.

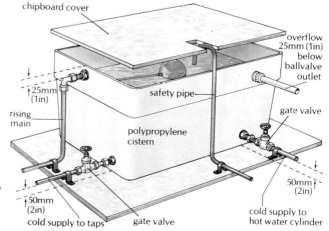

chipboard cover

overflow 25mm (1in) below ballvalve outlet

25mm (1in)

safety pipe

rising main

gate valve

polypropylene cistern

50mm (2in)

50mm (2in)

cold supply to taps

gate valve

cold supply to hot water cylinder

REPLACING A HOT WATER CYLINDER 61

Replacing a hot water storage cylinder is a job you'll be forced to do only if the existing one develops a leak; there is no alternative. Of course, you may want to put in a larger cylinder if the present one can't cope with the demands of a family first thing in the morning and again in the evening. If there are two adults and two children, for example, a cylinder with a capacity of 140 litres (30 gal) or 160 litres (35 gal) should be sufficient to prevent running out of hot water.

The other main reason for changing the cylinder is to install an indirect version in place of a direct type. It's certainly worth doing this if you live in a hard water area, as it prevents scale building up in the boiler.

However, before you jump in with your spanners, you've a little investigating to do to find out exactly how the existing system operates; it's not a bad idea to label the pipes next to the hot cylinder so you know where each goes. When you've decided whether you've got a typical direct or indirect cylinder, a self-priming indirect cylinder or a packaged hot and cold system (which could be direct, indirect or self-priming), you can decide on the type and size of cylinder you want to replace it and how you are going to deal with the pipework. Remember, if you want to install a new indirect cylinder you'll also need to fit a new feed and expansion cistern (header tank), unless you use a 'self-priming' system, which doesn't require one (see diagram). However, some plumbers don't like using these because they consider that the air bubble inside doesn't effectively separate the primary circuit from the domestic hot water.

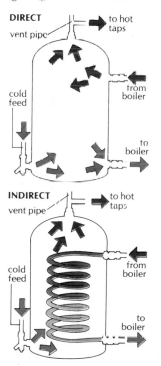

Types of hot water cylinder

With a direct cylinder (A), cold water from the cistern is fed into the bottom of the tank and then passes to the boiler for heating before being returned to the cylinder. In contrast, an indirect cylinder (B) contains a heat exchanger which heats the water in the cylinder. Self-priming cylinders (C) work on a similar principal, but the water in the heat exchanger is kept separate from the water in the cylinder by an airlock. You can also get a self-priming packaged system (D) which incorporates a cold water cistern on top to feed the hot cylinder below.

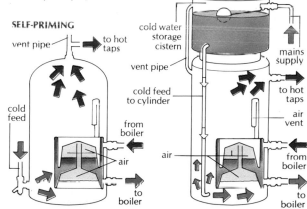

62 **REPLACING A HOT WATER CYLINDER**

The first, and obvious, thing to do is to turn off the boiler. Before you disconnect the immersion heater (if there is one) from its switch, you must turn off the electricity supply. Then you can drain the cylinder. Close the gatevalve on the cold water feed pipe and open the hot taps. Unfortunately, if there is no gatevalve, you'll have to tie up the ballvalve in the cold water cistern and then drain this, too. However, if you fit a rubber bung in the mouth of the outlet in the cistern, you can restore the cold water supply.

The hot water cylinder should have a draincock on its side near the bottom, so you can run the water out of this and through a hose feeding into a yard gully. Alternatively, if there is no draincock, you'll have to siphon the water out. To do this, disconnect the outlet on the top of the cylinder and fill a length of hose completely with water; put your thumbs firmly over the ends. Insert one end of the hose through the outlet and deeply into the cylinder, taking the other to a yard gully. Take your thumb off the end, and the cylinder should empty itself.

If you've already got an indirect cylinder, you will have to drain the primary circuit as well, and you can do this from the draincock by the side of the boiler. It's important that you allow the system to cool before attempting this. Don't forget to tie up the float arm of the ballvalve in the feed and expansion cistern, otherwise you'll never drain the circuit! You may be lucky and have a stopvalve on the cistern feed.

With the system drained, you can undo all the connections linking pipes to the cylinder. You may have to bend the pipes slightly so that you can work

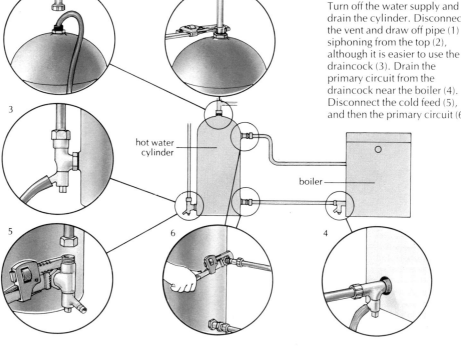

Removing an indirect cylinder
Turn off the water supply and drain the cylinder. Disconnect the vent and draw off pipe (1) if siphoning from the top (2), although it is easier to use the draincock (3). Drain the primary circuit from the draincock near the boiler (4). Disconnect the cold feed (5), and then the primary circuit (6).

hot water cylinder

boiler

REPLACING A HOT WATER CYLINDER 63

1 Set the cylinder on timber battens to allow air to get underneath. This will prevent condensation forming when cold water enters the cylinder.

2 Wrap PTFE tape round the connector that will link the feed and expansion pipe to the top of the cylinder. Straight versions are also available.

3 Screw the connector into the boss on top of the cylinder, making sure it is tight, and then join the pipework to it with a compression coupling.

4 Similarly, make up the male connector for the cold water inlet. Screw this into the female socket, making the final turns with a spanner.

5 Install a draincock just before the feed from the cold water cistern enters the bottom of the cylinder, so the hot water system can be drained.

6 The flow and return pipes to the boiler are joined to the cylinder with female/male connectors. Check these are tight in their sockets.

7 If fitting an immersion heater (see page 65), use the bosses provided in the top or side of the cylinder.

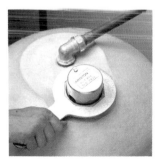

8 Use an immersion heater spanner to fix the unit firmly in place. Remove the cover to expose the wiring terminals.

9 Unless the cylinder is already insulated, fit a separate insulation blanket, leaving the heater control panel exposed.

64 REPLACING A HOT WATER CYLINDER

them free. This should enable you to drag the cylinder clear.

Before standing the new cylinder in place, check the 'tappings' of the inlets and outlets to make sure they haven't worked loose. Then, when in position, check that the tappings align with the pipework. It's quite likely that you'll have to adjust the runs, and often it's easier to do this by cutting back a short length of pipe and inserting a new section bent to the required angle.

You will certainly have to make some modifications if you're replacing a direct cylinder with an ordinary indirect one. The boiler flow and return pipes will have to be connected to the primary circuit heat exchanger in the cylinder. Just before the hot flow from the boiler enters the top of the heat exchanger, you'll have to 'tee' in an expansion pipe which leads up to a new feed and expansion cistern in the loft. The feed from this to the primary circuit is connected to the boiler return pipe taking water from the heat exchanger back to the boiler.

Now you can make up the cylinder connections and join the pipework. If you're not going to fit an immersion heater, you'll have to screw a blanking plate over the boss. With all this done, you can start to fill the system.

In order to prevent air locks, it's best to do this from the bottom up. If you are filling a direct cylinder or the primary circuit of an indirect cylinder run a hose from the cold tap over the kitchen sink to the draincock by the boiler. Make sure the plug is unscrewed and turn on the tap. The water entering will force the air back through the system. With a direct cylinder when water flows out of the vent pipe into the cold water cistern, you'll know the cistern is filled. Turn off the tap and close the draincock, then restore the cold feed to the cylinder either by turning on the gatevalve or by removing the bung

from the outlet in the cold water storage cistern.

With a primary circuit, when the water begins to flow through the feed outlet into the feed and expansion cistern this indicates that the system is filled. Turn off the tap, close the stopcock on or near to the boiler and then turn on the stopvalve controlling the flow of water into the cistern via a ballvalve, or release the float arm. If the boiler is connected to a central heating system, the radiator bleed valves should be left open when filling; close them as soon as the water begins to flow from them.

All that remains is to fill the rest of the cylinder, which you can do in exactly the same way as for a direct cylinder, connecting the hose to its draincock. Now, turn on the boiler or immersion heater to check the system; the expansion of the pipes caused by hot water flowing through them may weaken a joint, resulting in a leak. When you're happy that all is in order, you can lag the cylinder with a suitable proprietary jacket.

Refilling the system
By filling the system from the bottom through the draincocks, you'll reduce the risk of air locks forming. Connect one end of a hose to a tap supplied from the rising main, and the other end to the draincock. When the tap is turned on, the pressure of water will drive the air before it right back through the system.

draincock

hot water cylinder

IMMERSION HEATERS 65

Many people frown on using an immersion as a means of heating water, primarily because they think it is expensive to run. But if used wisely with a well-lagged storage cylinder, an immersion heater is an easy, clean and economical way of providing a hot water supply. It's also an ideal back-up system, particularly in summer when you may not want to go to the trouble, say, of lighting a solid fuel backboiler to get hot water.

An immersion heater fits into a special tapping usually provided in the cylinder, although you can cut your own access point. For the utmost economy, it's best to install a heater with a dual element or two separate units, one at the top of the cylinder, the other at the bottom. This way, if you only want a small amount of hot water you don't have to heat the whole cylinder; you simply switch on the upper element. Furthermore, if you connect the heater to a time switch, you can limit its operation automatically to periods just before peak demands are normally put on the hot water system.

If you are replacing a burnt out immersion heater, you should already have a suitable, nearby electricity supply to connect into. However, if

1 Wrap PTFE tape round the thread of the immersion heater, slip on a washer smeared with jointing compound and screw the heater into the boss.

2 Once the heater is in place, on some models, you will have to fit the thermostat in a separate operation. First remove the cover.

3 The thermostat consists of a control unit set on top of a long rod, which passes down through the casing into the cylinder.

4 Set the temperature to which you want the water heated, say 60°C, by turning the dial on the control unit with a screwdriver.

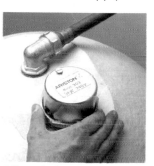

5 Wire the element to the thermostat, then connect 1.5mm² 3-core heat-resisting flex to the terminals as shown.

6 Connect the flex to a switched fused connection unit fitted with a 3 amp fuse and run on a 15 amp radial circuit.

66 **IMMERSION HEATERS**

you are putting one in for the first time, it's probably best to run a new 15A radial circuit from the consumer unit or fuse box to a 20A fused double-pole switch. Then, you can connect the heat-resisting flex from the control panel on the heater to this.

Alternatively, if you're installing a 3kW heater or one of a lower rating, you can plug the heater into a 13A socket on a ring or radial circuit. If you choose this method, consider installing a fused connection unit to supply it. This will ensure that the outlet point is only used for the heater so it cannot be disconnected inadvertently.

Again, don't attempt any electrical work unless you know exactly what you're doing.

Types of immersion heater
Probably the most common immersion heater is the type that protrudes down into the hot water cylinder, either vertically or at a slight angle, from a boss at the top. Sometimes these units run almost the full depth of the cylinder, and they are often used as a back-up hot water heating system, particularly to a coal-fired back boiler. A thermostat that is also inserted into the cylinder can be set to control the water temperature by switching the unit on and off.

However, this type of immersion heater has the costly disadvantage that all water in the cylinder is heated by expensive electricity, even though only a little may be required. To overcome this you can fit two smaller elements instead, one at the top, the other at the bottom, perhaps keeping the top one on permanently and only turning on the second to supply bath water. For this system to work efficiently the cylinder must be well lagged.

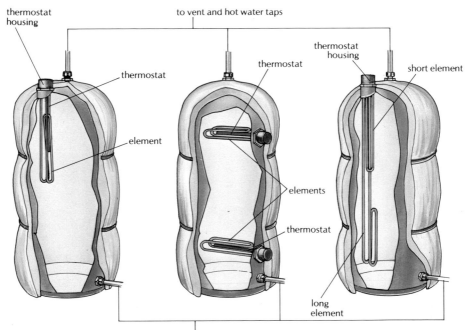

INSTANTANEOUS WATER HEATING 67

The commonest way of heating water for household needs is to have some form of hot water storage system with a boiler of some description that may also be powering a central heating system. You may have an immersion heater as well. However, there is an alternative, and that is some form of instantaneous water heating system that heats the water just before it flows out of the tap.

Gas systems are the most versatile of all. At the top of the range is a 'multipoint' heater which, as its name implies, will supply hot water to various outlet points round the house — over baths, basins and sinks, for example, and even showers. The heater is normally connected to the rising main and has to be sited against an outside wall because it has a balanced flue: the burners use air taken from the outside and not from the room where the heater is situated. You can also get small single point heaters to fit over a sink or basin.

Of the electric heaters, only the instantaneous showers operate in the same way as the gas heaters; that is they only heat the water when the tap is turned on. Other heaters operate like a very small storage system. Each comprises a small cylinder which incorporates two heating elements.

The smaller electric heaters come into their own when you need to provide water at an outlet point that would be difficult to supply by any other means. They are easy to install and they don't require an outside wall.

With an instantaneous gas water heater, when the cold water supply is turned on, water is conducted through a series of channels. These are heated by powerful burners which are ignited simultaneously. Some models also incorporate a cold tap so that cold water can be mixed with the hot on leaving the heater to give the temperature of water required.

Most of these heaters need a balanced flue (below) because the air for the burners has to be drawn from outside and not from the room where it is installed. Similarly, the exhaust gases must not be discharged into the room.

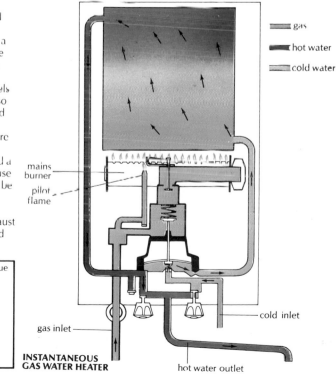

gas
hot water
cold water

mains burner
pilot flame
cold inlet
gas inlet
hot water outlet

INSTANTANEOUS GAS WATER HEATER

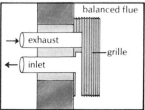

balanced flue
→ exhaust
← inlet
grille

68 A GARDEN WATER SUPPLY

When a house is built and the plumbing installed, it's rare that provision is made for an outside water supply. Considering that you need water for the garden and perhaps to wash a car, it's a pity at least the beginnings of a system — an outside tap fixed to the house wall — are not provided. Providing a garden water supply is one of the easiest plumbing projects you can carry out, but you must first get approval from your local water authority; you'll have to pay a higher water rate, and in some areas the garden supply may need to be metered.

Because an outside tap is likely to be used infrequently, take the supply from the rising main. Normally, it's most convenient to do this under the kitchen sink by teeing off the supply leading to the cold water tap. You then have to drill a hole through the outside wall, feed the pipe through and connect a tap to the other end. It's essential to install a stopvalve in the branch just after the tee so the pipe can be drained in winter — a vital safeguard that prevents water freezing in it and so damaging the run. Make sure that you use a suitable outside tap as well. The

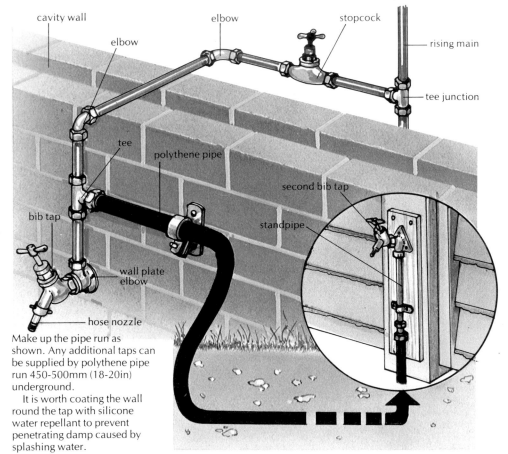

cavity wall

elbow

elbow

stopcock

rising main

tee junction

tee

polythene pipe

second bib tap

bib tap

standpipe

wall plate elbow

hose nozzle

Make up the pipe run as shown. Any additional taps can be supplied by polythene pipe run 450-500mm (18-20in) underground.

It is worth coating the wall round the tap with silicone water repellant to prevent penetrating damp caused by splashing water.

type to buy is an inclined bib tap, which screws into a backplate connector. The headgear of the tap slants outwards so that you can turn the simple crutch handle without grazing your knuckles on the wall. The tap should also have a removable hose union so a garden hose can be fitted easily.

Of course, the garden tap doesn't have to be set against the house wall. It could be positioned against a nearby garage wall, and even inside an outbuilding if needed to supply a washing machine. You could also install a standpipe further down the garden to

save having a hose snaking down the path ready for everyone to trip over it when in use.

To give the supply pipe as much protection as possible against frost and even the accidental blow from a spade, it's best to run it underground. For this, use either polythene or plastic pipe. The beauty of polythene is that you can buy it in long, coiled lengths so you shouldn't need any joins underground. Furthermore, the pipe connects easily to copper with compression fittings; it is not so susceptible to frost damage, and has good impact resistance, preventing damage by garden tools.

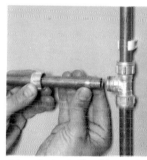

1 Insert a tee in the rising main above the main stopvalve after turning off the latter. Then connect in the 15mm branch to the garden tap.

2 Fit a stopvalve into the new run inside the house so the branch can be drained easily in winter to prevent freezing and when repairing the tap.

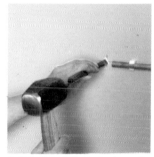

3 Chisel or drill a hole through the outside wall as near to the site of the tap as possible. Keep it small to reduce the amount of making good.

4 Use a 90° elbow to take the run down the outside wall to an angled tap connector incorporating a backplate.

5 Install an inclined bib tap fitted with a removable hose union. Wrap PTFE tape round the thread before fitting.

6 Use a 15 × 15 × 22mm tee to run a 12mm (½in) polythene pipe to other garden taps. This can be run underground.

70 DEALING WITH LEAKING TAPS AND PIPES

Signs
Tap fails to turn off properly.
Tap drips from spout, resulting in unsightly scale.

Cause Worn washer or worn seating.

Action — pillar and bib taps.
● turn off the water supply to the tap.
● unscrew and lift up the cover to expose the headgear nut.
● hold the tap's spout and undo the nut with a spanner.
● lift out the headgear and remove the jumper unit. The washer is usually attached to the bottom.
● undo the nut holding the washer in place, or prise the washer free.
● if the washer is difficult to remove, replace the jumper unit.

● bath taps require a ¾in washer, basin and sink taps a ½in washer, but you can turn the old washer over as a temporary measure.

If the washer is in good condition, the seating of the tap is worn:
● regrind the seating with a reseating tool (hire rather than buy one, but they are not that common).
● alternatively, fit a washer and reseating set.
● another easy solution is to fit a domed rubber washer that will push well down into the seating when the tap is closed.

After repair:
● replace the headgear and partially close the tap.

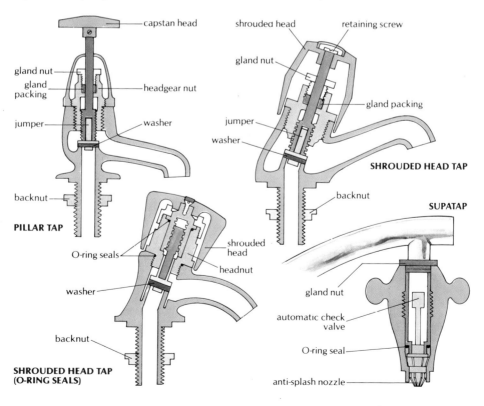

PILLAR TAP
— capstan head
gland nut
gland packing
headgear nut
jumper
washer
backnut

SHROUDED HEAD TAP
shrouded head
retaining screw
gland nut
gland packing
jumper
washer
backnut

SHROUDED HEAD TAP
(O-RING SEALS)
O-ring seals
washer
shrouded head
headnut
backnut

SUPATAP
gland nut
automatic check valve
O-ring seal
anti-splash nozzle

DEALING WITH LEAKING TAPS AND PIPES 71

● restore the water supply with the tap runing to prevent airlocks.

Action — shrouded head tap.
This type of tap is a more modern design than the previous taps, but the procedure for rewashering and reseating is similar. The biggest problem is in deciding how to remove the shrouded head itself.
● turn off the water supply to the tap.
● prise off the hot or cold indicator cap to reveal the retaining screw.
● undo the screw and remove the shrouded head.
(Alternatively, the retaining screw may be on the side of the head, or the head may just pull off.)
● hold the spout and release the headgear nut, then take out the headgear.
● undo the screw holding the washer in place and secure a new washer.
● reverse the procedure to reassemble the tap and restore the water supply

Action — Supatap.
With this design of tap there is no need to turn off the water supply.
● turn on the tap a little so there is an even flow of water.
● hold the nozzle in this position and undo the head-nut.
● turn on the tap fully and continue to unscrew the nozzle.
● a cut-off valve will stem the flow of water as the nozzle comes away.
● tap the nozzle upside down in your hand to release the anti-splash device; the washer is attached to its end.
● prise the washer unit free and fit a new one.
● reassemble the tap.

Signs
Water oozing round spindle, or from under shrouded head, with the tap fully closed.
Juddering in the plumbing system when

the tap is operated.
Tap turns on too easily.
Cause Worn gland packing on old taps, worn O-ring seals on new taps.

Action — old pillar and bib taps.
● remove the handle and cover from the tap.
● try tightening the gland packing nut.
● if this doesn't work, replace the packing round the headgear with graphite-impregnated string (above), which is specially made for the purpose. Make sure you pack it down well. You could use wool steeped in petroleum jelly as a stopgap measure.
● replace the packing nut.
● reassemble the tap and restore water supply.

Action — modern pillar and shrouded head taps with O-ring seals.
● turn off the water supply.
● remove the headgear (see above).
● remove the circlip on top of the headgear to reveal the seals.
● prise off the seals with a small screwdriver, but be careful not to damage their seatings.
● slip on seals of the correct size.
● reassemble the tap and restore water supply.

Action — Supatap.
If water seeps from round the top of the nozzle, you will have to fit (or replace)

72 DEALING WITH LEAKING TAPS AND PIPES

an O-ring seal underneath the head-nut. To do this, remove the nozzle as if you were rewashering.

Action — mixer taps.
Treat the hot and cold controls as shrouded head taps. The O-ring seals on a swivel spout may sometimes need replacing. The spout can be released either by turning it to one side and pulling it from its mounting, or by releasing a grubscrew.

Leaking pipes
Cause Frost, corrosion, mechanical failure resulting from a drill or nail being driven through the pipe.

Action
● if any electrical equipment is affected by the leaking water, or caused the leak in the first place, turn off the power supply immediately. Remove the offending equipment and check nearby ceiling roses and sockets

to make sure no water has seeped behind them. After making sure that all electrical accessories are dry, you can turn on the power again.
● turn off the water supply to the damaged pipe. Drain down that section.
● cut out the damaged section of pipe and insert a new short length. If it's only a small hole you're mending, you may only need a straight coupling.
● restore the water supply and check the repair for leaks.

Proprietary emergency repair kits are also available:
● plastic putty that has to be mixed with a hardener before being packed over the hole or into a leaking joint.
● sticky waterproof tape that has to be wound round the pipe or fitting.
At best, these products are a temporary measure. They are not 'instant' because you still have to drain the pipe. The most effective immediate repair is to use a pipe clamp.

One of the main advantages of the tape repair kit is that it can be used on leaking joints as well as punctured pipes. It is a two-part system consisting of a base tape (1), which is wound tightly round the pipe or fitting for 25mm to 35mm (1in to 1½in) each side of the leak. The second tape (2) is applied over this and continued for a further 25mm to 35mm (1in to 1½in) each side of the first tape, being stretched slightly as it is wound round the pipe.
  Where space permits on pipe runs, a pipe repair clamp (3) makes an effective seal. It is an 'instant' repair that does not need the pipe to be drained beforehand. But it is not suitable for use on leaking fittings. Clip the pad and metal plate together round the pipe and then tighten the wing nut (4) to stem the flow.

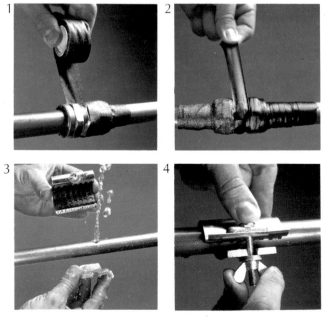

PROBLEMS WITH BALLVALVES AND WCS 73

Signs
Water continues to enter the storage cistern even though the float arm appears fully raised. The problem usually comes to light when water is seen discharging from the overflow pipe.

Action
● check the float at the end of the float arm. If it's perforated and water is seeping into it, the ballvalve won't cut off when the water reaches the correct level, resulting in an overflow. In this event, all you have to do is replace the float with a new one; simply unscrew the old float and screw on the new one.
● check the position of the float arm. By bending (or adjusting) it slightly downwards, you can apply more pressure to the valve mechanism to make sure it cuts off the supply of water before the cistern overflows.

If, after you've made these checks and adjustments, water still continues to flow into the cistern, the problem rests with the ballvalve itself. Use the information given below to identify the type fitted to your cistern and the repairs you may have to carry out.

Croydon pattern
This is the oldest type of ballvalve and is now fairly rare. Because it tends to be noisy in operation, you would be better off replacing it with a newer type rather than repairing it.

Portsmouth pattern
● first check the metal body for the initials stamped on the side:
HP = high pressure. This type should be used on main cold water storage cisterns.
LP = low pressure, fed from a storage cistern. This type should be used on WC cisterns and the like. If an LP valve is fitted to a main storage cistern, the valve will not shut off properly. There-

An equilibrium valve copes with mains pressure fluctuation by admitting water to a sealed chamber at the end of the valve. This ensures that an equal pressure bears on both ends of the sealing plug and only movement of the float arm will open it. A rubber disc stops the flow in a diaphragm valve, and a Torbeck valve combines facets of both designs.

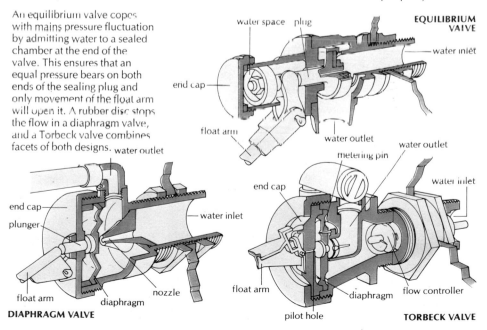

EQUILIBRIUM VALVE
water space   plug
water inlet
end cap
float arm
water outlet

DIAPHRAGM VALVE
water outlet
end cap
plunger
water inlet
float arm
nozzle
diaphragm

TORBECK VALVE
metering pin
water outlet
water inlet
end cap
float arm
diaphragm
flow controller
pilot hole

74 PROBLEMS WITH BALLVALVES AND WCS

fore, you should change the valve for the correct type.
● If the correct type is fitted, the problem is either a worn washer, a build up of scale or a corroded ball-valve seating. In the latter case, replace the valve.

To change the washer:
● turn off the water supply to the cistern.
● take out the split pin and remove the float arm.
● unscrew the retaining cap.
● insert a screwdriver blade into the slot where the float arm sits and push out the piston.
● unscrew the piston cap to release the washer. If the cap is difficult to undo, hold one end in a pair of pliers and hold the other under a hot tap. Alternatively, prise the cap round with a screwdriver.
● if you still can't remove the cap, use a nail to dig out the old washer and

work a new one into the seating.

If the valve is badly scaled:
● remove the valve by undoing the tap connector and the outer back-nut. This will allow you to withdraw the valve from the cistern.
● dismantle the valve and clean it with a wire brush and steel wool. Make sure the nozzle isn't obstructed by scale.
● smear the piston with petroleum jelly before reassembling.
● reassemble the valve.

Diaphragm valve (also known as the BRS or Garston pattern. Often, they have an overhead spray-type outlet for silent operation).

If this fails to cut off properly, check:
● the adjusting screw on the float arm which may be preventing it closing.
● the rubber diaphragm to see if it's been perforated. Sometimes, the diaphragm jams against the nozzle.

Rewashering a Portsmouth ballvalve
This is a job that can be done with the valve remaining attached to the cistern. The water supply, however, first needs to be shut down. If the valve controls the main cold water system, this means turning off the main stopvalve; if it is on a WC, close the gate valve on the supply pipe.
   Once you have removed the float arm by taking out the split pin, and have unscrewed the retaining cap, you should be able to lever out the piston (1). This needs to be dismantled (2) to give access to the washer (3). After pushing this out, a new one can be worked into the seating and the valve reassembled in the reverse order. Lubricate the piston with petroleum jelly (4) before replacing it.

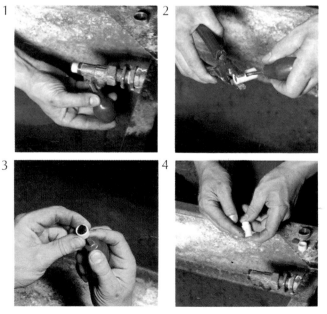

PROBLEMS WITH BALLVALVES AND WCS 75

● the nozzle to see that it isn't blocked by grit.

Noisy ballvalves

At one time, a silencer pipe, extending from the ballvalve outlet to below water level, was commonly fitted to reduce the gushing noise of the cistern filling up. However, because of the possibility of 'contaminated' water from the cistern being siphoned back into the mains supply, these devices are no longer permitted. So if you have a noisy valve, you'll have to consider replacing it with an equilibrium Portsmouth or diaphragm (Torbeck) pattern. This gives a quiet, fast refill, and also prevents water hammer.

A fluctuating water supply?

If the pressure in the mains supply varies with demand during the day, it could affect how well the ballvalve on the main cold water storage operates. To overcome the problem, fit an equilibrium valve.

Faulty WC cistern flushing mechanism
Signs

It takes several pulls in quick succession to start the water flowing through the siphon and into the pan.

Causes In a modern piston-type cistern, failure is due to worn siphon washers or flap valves covering the inlet holes in the base of the piston (see above right).

Action

● tie up the float arm to prevent water entering the cistern, then empty the cistern.
● disconnect the lever from the piston rod.
For a low-level cistern with flush pipe to pan:
● disconnect the flush pipe from the cistern by undoing the back-nut.
● undo the siphon retaining nut,

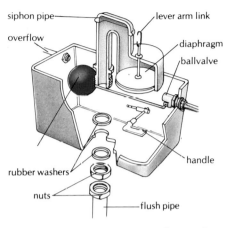

siphon pipe — lever arm link — overflow — diaphragm — ballvalve — handle — rubber washers — nuts — flush pipe

enabling you to remove the siphon from the cistern.

For a close-coupled suite:
● undo the retaining nuts on the underside of the cistern to release the siphon.
● separate the base of the piston from the siphon. You'll see the old flaps resting on top.
● replace these flaps with new ones, which are usually made of rubber or plastic that can be trimmed to size.
● make sure the new flaps move freely against the sides of the siphon (they shouldn't quite touch).
● reassemble the cistern in the reverse order.

Note

You may have an old Burlington pattern (bell-type) high-level cistern. In all probability, this will be nearing the end of its life. It is noisy in operation, and rust may be affecting its workings. Water continually entering the pan, even though the chain hasn't been pulled, is a sure sign of this. This type of cistern also attracts condensation. Rather than repair it, you would be better off installing a new modern type of cistern.

76 DEALING WITH BLOCKAGES

Signs
Water fails to flow from a sink, basin, bath or shower tray.

Action
● check that the outlet is not blocked by hair, soap, or kitchen waste.
If the outlet is unobstructed, the blockage is likely to be in the trap. This can be cleared either by using a plunger or by dismantling the trap.

Using a plunger:
● partially bale out the water.
● hold a damp rag over the overflow outlet.
● place the rubber or plastic cup of the plunger over the main outlet and work it up and down to force water through the trap

Dismantling the trap:
● an old lead trap will have an access eye at the bottom which can be unscrewed with a spanner (support the rest of the trap so that it doesn't buckle or twist).
● push a length of wire (an opened-out coat hanger will do) into both sides of the U-bend and rake out the debris. You can also push the wire through from the outlet.
● make sure you place a container under the trap to catch the water and blockage material.
● replace the eye and flush the system.

For a modern plastic P-trap:
● this usually unscrews in two places. Loosen both nuts, remove the bend and flush out under a tap. (Don't use the taps over the sink or basin you are working on.)
● replace the bend and flush the system.

For a modern plastic bottle or dip partition trap:
● unscrew the bulb and flush it out.

If you find that the trap is clear, the blockage has probably worked its way round into the outlet pipe. Usually, this is quite short so you can poke a length of wire along it to free the blockage before you replace the trap. On longer runs, check that the pipe doesn't sag, providing a collecting point for a blockage.

Signs
WC pan fails to clear after cistern is flushed, and water level is much higher than normal.

Cause
Blocked WC pan U-bend. Blocked main drainage run, causing a backlog in soil stack.

Action
● use a special plunger (which is similar to, but larger than the one used for sinks) to force the blockage through to the main drainage run.
● alternatively, try and work the blockage free, using a piece of wire.
If this fails, the blockage is likely to be in the soil stack or underground drainage run.

Note
A double trapped siphonic pan will block repeatedly if an object such as a full-length pencil or ballpoint pen gets into the second trap. To clear this, it may be necessary to remove the pan and use a string passed through the waterway to pull a ball of cloth through from the outlet to inlet.

Blockages can occur in the main drainage run at the foot of the soil stack, in the trap of a yard gully, or in the pipework and inspection chambers (manholes) linking the house to the main sewer. Don't forget to check rainwater hopper heads as well, as these sometimes become blocked with leaves.

DEALING WITH BLOCKAGES 77

To unblock a yard gully:
● use a trowel or a gloved hand to clear out the debris.
● rinse the gully — a jet of water from a hose will be far more effective than running in water from a tap.

If the drain is blocked at the foot of the soil stack or just a little way down from a yard gully, no waste will reach the first manhole on the run.
● check the manhole, and if the blockage is where you suspect, use drain clearing rods to free it (below).
● as these rods are fairly expensive, hire them instead of buying.
● screw the rods together to form a flexible pole, and attach the appropriate head to the front end.
● from the manhole, push the rod up the pipe towards the house to break down the blockage.
● always turn the rods clockwise so they don't come undone in the pipe. Retrieving them can be difficult.

● cover the outlet of the manhole so that when the debris from the blockage flows into the chamber it can be scooped out.

If the blockage is in the branch pipe just before it joins the main sewer, the manholes are likely to overflow.
● work at the lowest manhole on the branch. The outlet will be trapped to prevent sewer smells entering the chamber, and it may be this which is blocked.
● if the trap is clear remove the disc or plug covering the rodding eye (see diagram), allowing drain rods to be pushed into the pipe.
● by plunging and twisting the rods, you should be able to work the blockage so that it passes into the sewer.
● when the blockage has been removed, flush the system thoroughly. Make sure the rodding eye plug is refitted properly and replace the manhole cover.

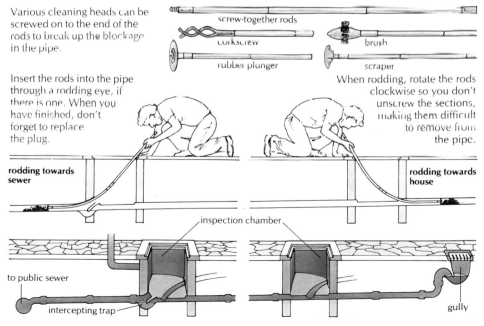

Various cleaning heads can be screwed on to the end of the rods to break up the blockage in the pipe.

screw-together rods

corkscrew   brush

rubber plunger   scraper

Insert the rods into the pipe through a rodding eye, if there is one. When you have finished, don't forget to replace the plug.

When rodding, rotate the rods clockwise so you don't unscrew the sections, making them difficult to remove from the pipe.

rodding towards sewer

rodding towards house

inspection chamber

to public sewer

intercepting trap

gully

78 PROBLEMS WITH GUTTERS

Signs
Gutter overflows during heavy rain.

Cause
Blockage in gutter due to build up of debris. Also sagging gutter run, or a blockage in the downpipe.

Action
Clear any debris into a bucket. If this isn't the cause, check for sagging with string lines between ends of gutter. Replace any loose brackets.

You may be able to scoop out a blockage at the top of a downpipe, but if it has passed into the eaves offset, or further down, you may have to dismantle the downpipe from the bottom up. Sometimes it's possible to ease the eaves offset from its sockets without disturbing the pipe.

Signs
Water leaking at gutter joint.

Cause
On cast iron and asbestos-cement guttering, this is probably due to the putty seal between sections of gutter having worn away. On PVC guttering, the sections may not have been clipped together tightly.

Action — cast iron and asbestos-cement guttering.
● unbolt joint (cut away rusted bolts).
● clean up contact faces.
● apply bed of mastic between faces.
● rebolt sections together.

Action — PVC gutters
● check that sections are fully clipped into connectors.

Signs
Water leaks through hole in gutter.

Cause
Rust in cast iron guttering, and particularly where ogee sections have been screwed to the fascia board.

Action
● Remove all traces of rust with a wire brush.
● Paint the affected areas with bitumastic paint.
● Use a proprietary repair kit to patch up hole or insert a length of plastic guttering.
(If gutters are generally in a poor state then it's worth replacing the entire system.)

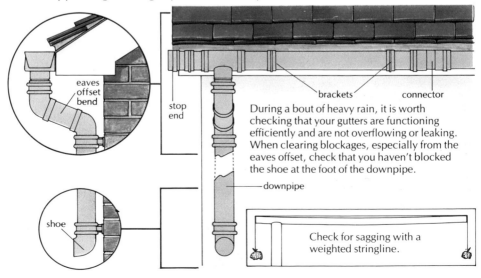

eaves offset bend

stop end

brackets

connector

During a bout of heavy rain, it is worth checking that your gutters are functioning efficiently and are not overflowing or leaking. When clearing blockages, especially from the eaves offset, check that you haven't blocked the shoe at the foot of the downpipe.

shoe

downpipe

Check for sagging with a weighted stringline.

Acknowledgements

Tools, Equipment & Facilities
The publishers are grateful to the many organisations and individuals who supplied materials, tools and other equipment, or provided locations and/or facilities for photography. Thanks are due especially to the following:
Barking-Grohe Ltd, Barking, Essex (Mr V H King): Thermostatic shower mixer and rose.
Bartol Plastics Ltd, Doncaster, Yorkshire (Trisha Brimblecombe): Plastics push-fit waste and Acorn hot-water systems..
City Electrical Factors Ltd, Stoke Newington, London (Mr Brian Vince): Immersion heater and electric shower unit.
Arthur Collier Ltd, Brixton, London (Mr John Collier): Plumbing tools.
Conex-Sanbra Ltd, Tipton, West Midlands (Mr John Pritchard): Compression fittings, stopcocks and valves.
Delta Capillary Products Ltd, Dundee, Scotland (Mr R McCutcheon): End-feed capillary fittings.
DHC Products (Denton Hat Co), Denton, Manchester (Mr J Greenhough): Pipe-leak repair clamp.
DRG Sellotape Products Ltd, Borehamwood, Herts (Mr J Wallis/Mrs J M Britton): Pipe and hose repair bandage.
Econa Appliances Ltd, Solihull, West Midlands (Elizabeth King, Link Communications): Parkamatic Silver waste-disposal unit.
Footprint Tools Ltd, Hollis Croft, Sheffield (Mr Christopher Jewitt): General purpose tools.
Houseman (Burnham) Ltd, Burnham, Slough (Mr Mike Trim): Permutit water softener and installation kit.
Hunter Building Products Ltd, Abbey Wood, London (Mr J P Ward-Turner): Genova solvent-weld hot and cold water system and bib tap.
IAZ International (UK) Ltd, Reading, Berkshire (Lesley Hawthorne/Megan Nixon): Zanussi automatic washing machine.
IMI Range Ltd, Stalybridge, Cheshire (Mr Ken Peacock): Supercal insulated hot-water cylinder.
Key Terrain Ltd, Maidstone, Kent (Mr H W Marden-Ranger): Terrain solvent-weld waste system and fittings.
Kitchen Design and Advice Ltd, Hendon, London and Enfield, Middlesex (Mrs Barbara Ellison/Mr Arlen Whittock): Kitchen base unit and sink top.
Monument Tools Ltd, Balham, London (Mr John Collier): Plumbing tools.
Royal Bathrooms Ltd, Stoke-on-Trent, Staffs (Ruth Francis, Paul Winner Market Communications Ltd/Mr Martin Morris): Bathroom suite, taps, shower tray and cubicle.
John Sydney Ltd, Walworth, London (Mr B Maizner): Cascade mixer tap.
Yorkshire Imperial Fittings Ltd, Leeds, Yorkshire (Mr Stephen Young): 'Yorkshire' capillary fittings.
Tools and equipment co-ordinator: Mike Trier.

Illustrations
The photographs and artwork in this book were specially commissioned from the following, to whom the publishers extend their thanks:
Photography **Jon Bouchier** 16-17, 22, 30-31, 33 (above), 35-59, 69, 74; **Simon de Courcy Wheeler** 14-15, 18-21, 23-29, 35, 63, 65, 72.
Artwork **Trevor Lawrence** 20-21, 39, 68; **Ian Stephen** 33; **Brian Watson/Linden Artists** 6-13, 26-32, 34-38, 40-67, 70-78.